# HAHNEMANN UNIVERSITY LIBRARY

# Radiologic Interpretation of ERCP

# Radiologic Interpretation of
# ERCP

## A Clinical Atlas

**Errol M. Bellon, M.D.**

*Professor of Radiology*
Case Western Reserve University
School of Medicine
*Director,* Department of Radiology
Cuyahoga County Hospital
*Formerly Chief,* Radiology Service
Veterans Administration Hospital
Cleveland, Ohio

MEDICAL EXAMINATION PUBLISHING CO., INC.
an Excerpta Medica company

Bellon, Errol M.
    Radiologic interpretation of endoscopic retrograde
cholangiopancreatography (ERCP).

    Bibliography: p.
    Includes index.
    1. Biliary tract—Radiography—Atlases.
2. Pancreatic duct—Radiography—Atlases.  3. Endoscope
and endoscopy—Radiography—Atlases.  I. Title.
[DNLM: 1. Cholangiopancreatography, Endoscopic
retrograde—Atlases.  WI 17 B447r]
RC847. B43    1983    616.3'650757    83-7892
ISBN 0-87488-707-0

Printed in the United States of America

*Science is our century's art.*

Horace F. Judson
*The Search for Solutions*

# Contents

# Preface

This book is a topography of the manifestations of pancreatic disease demonstrated on endoscopic retrograde cholangiopancreatography (ERCP). Thus, it intends to serve the clinician in either private or academic practice as a reference tool for those interesting, different, or difficult cases typically encountered. The radiographs provided are usefully seen as a set of template-like visual forms illustrating the variations of normal ductal radiographic anatomy through the range of typical (and a sampling of atypical) abnormal variations. It should be noted that the radiographs and their descriptions alone do not attempt to represent the complete map of this terrain. The clinical history and comment given with each case are as important to a thorough understanding as the visual components. In practice, the precise stage of assessment at which a clinical history becomes pertinent to diagnosis is variable, but that it assists in diagnosis is irrefutable. To some extent, the comment section of the cases presented is textbook discussion. However, it also replicates the kind of thought processes that occur, abstractly and in a kind of shorthand, as the radiologist studies the various projections. Thus, comment in each case considers those factors which, although not directly affecting the case at hand, are pertinent to like cases.

The criterion for inclusion here is that a given case be of teaching value insofar as one or more of the illustrative components are concerned. The range of variation shown demonstrates not only the complexity of pancreatic disease expression, but the sophistication of the technique of ERCP at its current stage of development as well. Thus, formula approach, except for a few broad generalizations, is inadequate to the task of interpretation.

The book is designed around a series of individual cases, further organized into five sections, each with a common theme. Each case is complete in itself, and generally concerns only a single patient; however, in Section VI (Self-Evaluation), the cases are sometimes composites of several patients, which serve to highlight similarities and relationships.

A uniform structure has been employed for each case, consisting of clinical history, radiograph(s), radiographic findings, diagnosis, and comment. In this way, the material presented may be seen as closely approximating the situation in everyday clinical practice. In the Self-Evaluation section the reader is confronted with the clinical history and radiograph(s) alone, and is invited to construct the remaining elements for himself as an exercise in self-learning. Only at this stage should he turn the page and compare his efforts with the findings, diagnosis, and comment, as well as referring to similar cases in preceding sections.

The sequence of arrangement of the sections is deliberate, and proceeds from the simplest (normal ERCP), through moderately involved (chronic pancreatitis), to most complex (carcinoma). In this way, the reader may use the sections as building-blocks. The preceding notwithstanding, the reader is encouraged to use the book on an *ad hoc* basis, to skim, and to refer to individual cases as needed, using the index as his guide.

A note on nomenclature: in relation to the pancreatic duct, the terms "proximal" and "distal" refer to the segments occurring, respectively, in the tail or the head of the pancreas. This usage accords with the direction of flow of secretions within the pancreatic duct.

Many hands are involved in this work. My thanks to the technologists in the Department of Radiology, whose skill, at times art, is responsible for the quality of radiography illustrated on these pages; and to my colleagues in the Department who have shared in nearly all of the individual examinations, particularly Dr. Ali Jumshyd. I am, as well, appreciative of the collaborative efforts of Dr. John Marshall, of the Division of Gastroenterology, who was the endoscopist for nearly all the patients we have examined. Dr. Mary Petrelli generously consulted on the pathology of individual cases and provided, as well, the illustrations for Cases 30 and 31.

Medical editing is a science in its own right, invaluable as it illuminates and shapes the text for, merely, clarity. Ms. Sharon Lillevig has been involved in this book from its inception, and, indeed, contributed to its design. Mr. Donald Abbott, formerly of Medical Examination Publishing Company, believed in the work and sponsored it in many ways, as has Ms. Janice Resnick.

# Notice

The author and the publisher of this book have made every effort to ensure that all therapeutic modalities that are recommended are in accordance with accepted standards at the time of publication.

The drugs specified within this book may not have specific approval by the Food and Drug Administration in regard to the indications and dosages that are recommended by the author. The manufacturer's package insert is the best source of current prescribing information.

# I

# General Introduction

# HISTORY

As is the case with most new techniques, the path to ERCP (endoscopic retrograde cholangiopancreatography) pays tribute to a succession of pioneers. Thus, successful peroral cannulation and demonstration of the pancreatic duct rests on a history that begins with its use in anatomic studies [Pillan 1909, quoted by Millbourn[13], Rienhoff[18]], which led to its occasional use as a surgical adjunct by 1951 [Doubilet and Mulholland[6]]. By 1961, Anacker[1] and by 1965, Sarles[20], had constructed an anatomic basis for pancreatographic diagnosis.

Credit for the first nonoperative in vivo cannulation belongs to Rabinov and Simon[17], who succeeded in blind cannulation of the pancreatic duct using fluoroscopy to guide a steerable cannula/coil-spring instrument which "poised" a cannula over the usual anatomic location of the papilla. The procedure succeeded in one of eight cases, demonstrating the pancreatic duct and a dilated common bile duct with retained calculi. In 1967, Waldron[23] described a technique of experimental reflux pancreatography.

Adoption of the direct cannulation technique as a practical matter was made possible by the application of the new technique of fiberoptic endoscopy, which itself rested on the development of fiberoptics, specifically that of the gastroduodenal fiberoptic endoscope by Hirschowitz in 1958[7]. In 1966, Watson[24] described direct vision of the ampulla through a gastroduodenal fiberscope. Credit for successfully cannulating the ampulla under direct vision and radiographically visualizing the pancreatic duct belongs to McCune, Shorb, and Moscovitz[12], who achieved this combination in 25% of 50 cases. They modified an Eder duodenal fiberscope by incorporating a balloon at the tip of the movable catheter. Oi[15] developed a new fiberduodenoscope expressly for pancreatography. This was shortly followed by use of the technique for cholangiography as well as pancreatography by Oi[16] and by Takagi[21].

Millbourn[13,14] performed anatomicoradiologic studies on 150 autopsy cases and defined normal and abnormal appearances of the pancreatic duct. Dawson[4] and Kreel[10,11] carried out similar autopsy studies, and the latter gave details of the normal appearance of the pancreatic duct and also described variations due to aging.

The radiologic appearance of ERCP was characterized by Takagi[22], Kasugai[8,9], Classen[3], Demling[5], and Rohrmann[19]. The procedure has remained essentially as developed in 1974, such that by 1976 Bilbao[2] was able to define the complications in a series of 10,000 patients.

3

## TECHNIQUE

The primary focus of ERCP for the radiologist is on meticulous technique, in terms of the production of quality radiographs. Because diagnosis of pancreatic duct abnormalities is critically dependent upon definition of fine structural detail, ERCP demands technique of the highest quality.

By accepted standards of practice, quality ERCP is performed by a team comprised of the endoscopist and radiologist wherein the radiologist's primary functions, aside from film interpretation, include a) positioning of the patient; b) timing, selection of, and volume of contrast material injected; and c) selection of optimum projection(s) and timing of radiography.

### Cannulation

While the endoscopist and radiologist work in close collaboration during ERCP, the technique of successful cannulation is considered in such great detail in the literature on endoscopy that it will not be taken up here. Suffice to say that the technique is highly functional in the hands of a competent endoscopist, and successful cannulation is regularly achieved in more than 95% of the cases. In that regard, failure of cannulation of the ampulla in experienced hands is considered demonstration of possible significant structural or pathologic abnormality of the duct or ampulla.

### Injection

With the catheter advanced 1–2 cm into the pancreatic duct, hand injection of a water soluble contrast material is commenced. While 50% Hypaque, 60% Urografin[4,6,8], and metrizamide and metrizoate[5] have been used, 60% Renografin seems an optimal choice as this material is sufficiently dense to permit resolution of fine structure, while not so dense as to obscure small calculi or the like, nor too viscous to inject readily. For cholangiography, especially where the ducts are dilated, the Renografin is diluted to 30% concentration, so as not to obscure calculi and other small detail, which may be the case when the contrast material is too concentrated.

There is no fixed volume of contrast material which can be recommended for injection. Generally speaking, the accepted criterion for adequacy of examination is complete delineation of the entire pancreatic duct shown in whatever projections are necessary to achieve that goal. In practice, during the procedure one should expect opacification of first the main duct through to the tail (or to a point of obstruction), then of the major side branches. Fluoroscopic visualization of the minor side branches should not be attempted, nor should one see evidence of acinarization, both of which indicate that excessive contrast material has been injected. Should either of these be visualized, injection should be terminated.

In order to achieve optimal visualization without complication, the contrast material is injected at 1–2 ml increments by hand with fluoroscopic monitoring continued throughout the procedure for signs of any overfill. Gentle, continuous pressure on the syringe is wanted rather than rapid, intermittent maximal or harsh pressure. Pressure of up to 100 lb/in$^2$ can be generated within a syringe during injection even with moderate, sustained levels of pressure. However, that technique, in combination with contrast material viscosity and narrow catheter bore, combine to reduce intraductal pressures to safe levels. Furthermore, in relation to actual technique, one is dealing with an "open system" in that the endoscopic catheter does not completely occlude the pancreatic duct and, thus, permits backflow of contrast material. For this reason excessive pressure buildup cannot occur provided one does not a) wedge the catheter,* or b) inject in sudden, forceful spurts.

Total volume injected depends upon the capacity of the duct system and the amount of backflow, and varies by individual case. While the capacity of the normal pancreatic duct is about 2–5 ml and of the biliary tract 10–15 ml, a dilated biliary system may require 50–75 ml. In an effort to develop a standardized technique of injection, a method of manometric measurement and control of injection pressure during ERCP has been attempted [Kasugai[3]]. However, this method does not significantly alter quality of either procedure or results. The best gauge for volume and pressure of contrast material injection during ERCP remains, for all cases, adequate opacification.

## Filming

Radiographic spot films are obtained in sufficient quantity and at such projections as to completely delineate the duct from the ampulla to the tail of the pancreas. Occasionally, three, two, or even one view might suffice, but usually a series of projections is obtained, unfolding and laying out each kink and twist of the duct system.

At the conclusion of spot filming, a single overhead radiograph is obtained of the upper abdomen, patient supine, intending to provide the highest possible resolution of fine duct structure. In cases with suspected obstruction, or demonstrated stenosis, an additional AP radiograph is obtained at 5 minutes, and another at 10 minutes. These are all consequent on removal of the endoscope and catheter.

## Positioning/Filling

It is useful to consider the orientation of the pancreatic duct since its filling is gravity dependent. The course of the pancreatic duct in its distal one-third is a gentle anterior curve as it passes from the second portion of the duodenum to cross anterior to the lumbar spine; and then there is an even more gentle curvature posteriorly in the proximal one-third as it reaches the tail of the pancreas. This point is best appreciated by referring to a CT (computerized tomography) scan (Fig. I) taken through the plane of the pancreas, where the cross-sectional relationship of these structures is readily appreciated.

The examination is conducted with the patient in the prone, right posterior oblique (RPO) position, that is, with the right shoulder and hip elevated between 30° and 45°. The left shoulder and hip are in contact with the radiographic table.

---

*For exception to this rule see Case 46.

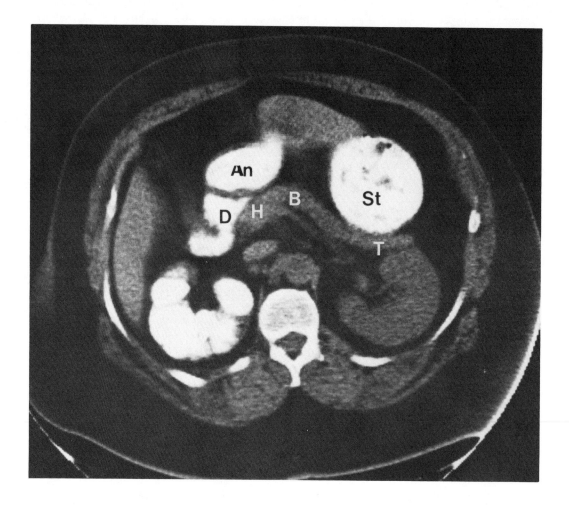

**Figure I.** CT scan, transverse section through the level of the pancreas. The head of the pancreas (H) lies anterior to the inferior vena cava, and medial to the second portion of the duodenum (D). The body of the pancreas (B) is located about 5 cm more anteriorly, while the tail (T) is located at a level posterior to that of the head. Contrast material also outlines the stomach (St) and antrum (An). A pancreas with a more transverse orientation than is commonly seen is presented here to better demonstrate the anatomical relationships. Usually, the body and tail extend obliquely in a cephalad direction, so that the tail of the pancreas is located near the hilum of the spleen. The extent of anterior convexity seen here, however, is normal.

During the procedure the effect of gravity is employed to assist in ensuring complete filling of the pancreatic duct as, for example, the initial RPO projection directs the flow of contrast material toward the tail. In addition, the radiographic table is tilted to a 30° or 40° Trendelenburg position (head down) since the pancreas is normally somewhat oblique upwards with the tail more cephalad than the ampulla.

To visualize the biliary tract, the endoscopic catheter is directed along the plane of the common duct so that it can be independently cannulated. The prone RPO position generally results in satisfactory filling of the left hepatic duct branches. The right lateral projection, assisted by the Trendelenburg position, is useful for visualizing the right hepatic duct branches and also for adequately filling the gallbladder and the cystic duct.

### Complications

The chief complications of ERCP are elevation of serum amylase after ERP and biliary infection after ERC. Transient elevation of amylase levels after ERP is common, and returns to normal in a majority of cases within a short time and in the vast majority of cases is asymptomatic. Ileus lasting during the day of the examination, and simple intramural deposition of contrast material also occur commonly and are not considered here in the category of complications. The following discussion of Bilbao's findings[1] on complications in a series of 10,000 patients demonstrates the quality of ERCP performance since there is documentation of a decreasing rate of complication (from 15% to 3.5%), as operator skill increases. The specific complications presented here represent all of those which occur with any significant frequency, and for each the percentage of incidence, according to Bilbao*, is given.

**Pancreatic Sepsis/Pseudocyst Abscess.** These have an incidence of 0.3% and constitute a serious complication responsible for extreme morbidity with prolonged hospitalization and death in 20% of cases, and are nearly always associated with cannulation and injection into an obstructed duct. This accounts for one-third (8 of 25) of the mortality in this series.

Tumor as a cause of the obstruction predominates over calculus (5 cases versus 1). The combination of tumor obstruction with sepsis accounted for 4 of 5 fatalities in this group.

The presence of a pseudocyst is *not* a contraindication to ERCP: 91% of ERCP's in patients with pseudocyst were free of complications (see Cases 24, 25, 26, 27, and 42).

---

*An interesting additional study was performed by Takemoto[7], who tabulated the results of a rettrospective study in 120 Japanese hospitals, covering 21,181 procedures. Complications were recognized in 1.2%, and the mortality was 0.12%. The cause of death was biliary infection (9 cases), pancreatic infection (5 cases), peritonitis (4 cases), perforation (2 cases), and bleeding (2 cases).

**Instrumental Injury.** Instrumental injury accounts for an incidence of 0.2%. Two of the 16 cases in this category were fatal. Perforation and intramural injection occur and are due to confusing or distorted anatomy, abnormal duodenal anatomy, and cannulation failure. Commonly associated conditions are carcinoma of the ampulla or pancreatic head, and previous GI tract surgery.

**Injection Pancreatitis.** Injection pancreatitis had an incidence of 1%, with no mortality. It represents the most frequent complication. Sixty-six percent of these had had pancreatitis previously. In 45%, acinar filling in various degrees was visible during fluoroscopy or on the radiographs.

**Cholangitic Sepsis.** This complication has an incidence of 0.8%, with a mortality of 10% of these. It represents the second most common complication and manifests clinically as an acute, febrile, often septic illness developing within 72 hours. It is the most common cause of death of the various complications of ERCP. Ninety-two percent of the incidence is accounted for by ERC. All eight deaths were in patients with obstructed ducts. Although calculi and tumor were equally often the cause of obstruction, tumor was the cause of the obstruction in all but one of the fatal cases (7 of 8). This can be prevented through prompt surgical relief and decompression after diagnosis. Gregg[2] noted that bile in a nonobstructed system is sterile. He observed positive cultures only in systems with partial or complete obstruction.

**Other Complications.** These include, more rarely, drug reaction (0.6%) and aspiration pneumonia (0.1%).

## INDICATIONS FOR ERCP

In considering the indications for ERCP it is useful to bear in mind that the pancreatic and bile ducts reflect abnormalities within their lumen, in their walls, or extrinsically, in the parenchyma adjacent; and that, regardless of pathophysiology, there is not only a finite limitation to the radiologic expression of these abnormalities, but a narrow one as well. Thus, for radiologic purposes, the pertinence of an obstruction of a duct which results from intrinsic tumor is its appearance—the "cutoff," the lobulated margins—not the histologic type of tumor expressed.

Throughout the pancreas, the focus of ERCP is on the etiological, concrete phenomenon (calculus, diverticulum, etc.) as the key to *indication* as such rather than the disease process. In the most simplified terms one considers whether one can anticipate visualizing calculi, not whether the patient suffers from cholelithiasis per se. Thus, at its most basic level, the intent of ERCP is discovery and identification of the products of disease. After ERCP is performed, the concrete findings discovered and identified will be related, in anatomic and pathophysiologic terms, to the clinical setting. Synthesis of all moieties leads, as a final step, to diagnosis. Radiologic interpretation includes all of these steps.

The indications for ERCP follow a pragmatic approach that is clinically oriented and directed toward conditions which may be seen, relatively speaking, as commonplace. Rare conditions which have been described in the literature as "findings" of ERCP (e.g., sclerosing cholangitis) are, in fact, *encountered* in the course of a study initiated without intention to search specifically for the problem which may have been discovered. Thus, such rare cases as noted previously (see also Cases 28, 43, 46, 47, and 48) are inadvertent and not to be perceived as functionally part of the procedure intent, or the differential, nor should they determine the choice/selection of diagnostic modality.

Furthermore, because ERCP is invasive, uncomfortable for the patient, and requires a high level of operator mastery, it has intentionally been used exclusively for those cases where yield will directly, significantly affect therapy.

Indications fall into few, fairly well-defined categories*:

1. Patients with known pancreatitis, to establish the cause and exclude a surgically remediable condition.

2. Patients with known pancreatic abnormality, to characterize the lesion (carcinoma versus pancreatitis).

3. Patients with undiagnosed abdominal pain, to establish whether the pancreas is the cause.

4. Patients with jaundice, to diagnose cause and to define structural changes in the pancreatic and bile ducts.

5. Patients presenting with metastatic tumor, to search for an obscure primary lesion.

## Patients With Known Pancreatitis, Cause Undetermined

Calculus, duct stenosis, papillary tumors, and diverticula are readily identifiable. ERCP is the only method for showing congenital abnormalities of duct architecture. In relation to contemplated surgical drainage procedures, it shows involved and uninvolved segments and sites of stenosis and demonstrates pseudocysts as well.

## Patients With Known Pancreatic Abnormality

An abnormality of the pancreas has been detected by some other method, and ERCP is performed to refine the initial characterization of the pathology. Primarily, these include cases where abnormalities are detected by other radiologic modalities, especially by ultrasound, CT, or barium upper gastrointestinal series; or the abnormality is noted on clinical examination as a palpable mass.

---

*For an excellent discussion on the clinical utility of ERCP see Cotton, P: Endoscopic Retrograde Pancreatography in *The Exocrine Pancreas,* (eds) Howat and Sarles. Saunders, London, 1980, pp. 293-295.

## Patients With Undiagnosed Abdominal Pain

Because of the essential need to distinguish/identify resectable from nonresectable forms of carcinoma of the pancreas, for the patient where pain is the sole presenting symptom, ERCP gives critical information about the state of the papilla, stomach, and duodenum as well as about the pancreatic duct. A normal study almost entirely excludes pancreatic neoplasm.

## Patients With Jaundice

In conjunction with ultrasound, CT, and percutaneous transhepatic cholangiography (PTHC), ERCP gives specific information as to the site, and frequently the nature, of obstruction or stenosis of the pancreatic or biliary ducts. In practice, ultrasound and CT are usually the initial procedures performed, followed by ERCP (with or without PTHC).

In those cases where the extrahepatic biliary system is obstructed or there is indication of a mass in the pancreas, ERCP indicates exactly the site and level of lesion, frequently the cause by implication. It, thereby, permits preoperative assessment of choice of surgical procedure.

## Patients With Unknown Primary Tumor

Although ultrasound and CT are highly accurate in excluding pancreatic neoplasm, there exists a small group of patients wherein the findings of both are negative or equivocal, and ERCP is used as the final arbiter of pancreatic normalcy. The practical assumption has been that a tumor within the pancreatic parenchyma will be reflected on ERCP as an identifiable abnormality of the duct itself *when such a tumor has reached the stage where it is symptomatic.*

## SELECTION AND CHOICE OF RADIOLOGIC PROCEDURE

Today, the radiologic studies available for examination of the pancreas reflect the considerable advances in diagnostic imaging capability during the past decade. Chiefly, the procedures include ultrasound, CT, ERCP, PTHC, percutaneous thin-needle aspiration biopsy, and arteriography. Nuclear imaging of the pancreas is now seldom employed because of its low specificity [Barkin[1], DiMagno[2]]. Barium studies, such as upper gastrointestinal series, are of very limited value, and chiefly are of help in detecting large masses in the head of the pancreas [Lukes[11]]. Conventional radiography and oral cholecystography/intravenous cholangiography have a narrow application, the latter being severely limited in application by the presence of jaundice. Pancreatic and mesenteric arteriography and venography are rarely necessary, and are used only occasionally for diagnosis of neoplasm, especially functioning tumors, and for preoperative assessment of operability [Freeny[6], Suzuki[15]].

## Diagnosis of Tumor or Cyst

For problems where the clinical history suggests tumor or cyst, ultrasound and CT are the procedures of choice. Ultrasound is limited in practice by the presence of bowel gas, obesity, and recent surgery. The incision, drains, and dressings interfere with good transducer/skin contact and the failure rate may be as high as 25%. However, it is less expensive and generally more accessible than CT, which has greater resolution capability and, therefore, can demonstrate smaller lesions (1–2 mm diameter).

CT is especially advantageous for display of tumor topography [Foley[5]], which has led to the establishment of diagnostic criteria of tumor resectability [Freeny[7]]. CT is not as much affected by the presence of obesity and bowel gas as is ultrasound, but is sensitive to artifact resulting from patient motion. The CT installation is fixed in site, while ultrasound equipment is portable and can be brought to the bedside of a patient too ill to be moved to the CT scanner.

In clinical studies, both ultrasound and CT have demonstrated accuracy in excess of 95% [Ralls[13], Freeny[7]] so that either may serve as the initial procedure for diagnosis of mass or cyst. Choice of one over the other is influenced by factors already described as well as by availability of equipment and individual expertise.

Given the predictive value of a negative examination at about 90% for ultrasound and 99% for CT [Freeny[7]], further studies are generally unnecessary assuming one or both of these are normal. ERCP is reserved for those cases where they are technically unsatisfactory or equivocal, or where a strong clinical suspicion of malignancy persists. There is, thus, a complementary role for ERCP used together with either CT [Moss[12]] or ultrasound. There is an additional role for ERCP, shared with PTHC, in defining the site and extent of biliary obstruction. Here, PTHC has the advantage of providing access for placement of biliary drainage catheters [Ferrucci[4], Hansson[9], Ring[14]]. Finally, percutaneous pancreatic thin-needle aspiration biopsy is increasingly used to verify the CT or ultrasound diagnosis of malignancy [Wittenberg, Yamanaka].

## Diagnosis of Pancreatitis

For pancreatitis and its complications, conventional radiography demonstrates pancreatic parenchymal calcification, duct calculus, incidental cholelithiasis, and the rare calcified functioning adenoma. For cholelithiasis, oral cholecystography may also be used in the nonjaundiced patient and, along with intravenous cholangiography, is capable of demonstrating the common bile duct.

Ultrasound is noninvasive and defines the pancreatic parenchyma well. It demonstrates the gland's size and shape, and shows textural changes associated with acute and chronic pancreatitis. It clearly demonstrates duct dilatation, both biliary and pancreatic. However, up to 50% of patients with chronic pancreatitis in remission may have a normal ultrasound scan [Gowland[8]] and many patients with mild degrees of acute pancreatitis show no radiologic abnormality, whether studied by ultrasound, CT, or ERCP.

CT appears to have advantages over both ultrasound and ERCP in detecting the changes of chronic pancreatitis, such as small intrapancreatic pseudocysts, pancreatic calcification, and fascial thickening [Foley[5], Lawson[10], Ferrucci[3,4]]. It has limitations similar to those of ultrasound regarding mild degrees of acute pancreatitis or chronic pancreatitis in remission.

ERCP provides more detailed information than either ultrasound or CT on pancreatic duct morphology, calculi, stenoses, and dilatations. It is especially helpful in diagnosing minimal degrees of pancreatitis (Cases 17 and 19)* and for showing localized sites of obstruction associated with such complications of pancreatitis as pseudocyst or calculus (Cases 21, 23–27 and 42).

ERCP is avoided in the presence of active pancreatitis because of the increased incidence of complications associated with the procedure, and also in known cases of pseudocyst or abscess, unless immediate surgical intervention is contemplated.

## Diagnosis of Bile Duct and Hepatic Abnormalities

ERCP is not primary to diagnosis of abnormalities of the liver and bile ducts. Rather, it is reserved for definitive evaluation when other methods such as oral cholecystography, intravenous cholangiography, PTHC, radionuclide scanning, etc., are equivocal, or define an abnormality without precise delineation. ERCP, specifically, identifies calculi, tumor, cysts, stricture, and extrinsic compression, much as it does with the pancreatic duct (see Cases 44C, 48, and 49).

# RADIOLOGIC INTERPRETATION

As has been mentioned earlier, there is a sequence of events involved in radiologic interpretation which includes *discovery* and *identification* of concrete findings, *assessment* of normal versus abnormal entities, and finally, *synthesis* of all these elements to produce a single diagnosis.

For ERCP, there are no special requirements of interpretation for the radiologist that do not apply equally to all other branches of radiology. For example, it is as necessary to distinguish narrowing, stenosis, and stricture of the descending colon as demonstrated by barium enema examination, as it is to do so with the pancreatic duct on ERCP. Depending on pattern recognition alone is as inherently risky with cholecystography as with ERCP.** On the radiograph a calculus, for instance, is not recognizable as such but as it produces a filling defect against a contrast material background.

If there is a rule of thumb for radiologic interpretation of ERCP, it is to keep in mind that, for the few diseases which manifest in the pancreatic duct itself, there is remarkable range of variation in expression. Good diagnostic technique,

---

*Even with mild cases of pancreatitis, especially during remissions of the disease, the ERCP may be normal, but, since the frequency with which that occurs is not clearly established owing to the difficulty of histologic confirmation, ERCP seems warranted.

**However, the reader will find excellent and useful examples of characteristic shapes and patterns illustrated in sketch form in the literature by these authors: Stewart[9], Ohto[4], Rohrmann[7], and Takemoto[10].

therefore, demands far more individual attention to fine detail than such a narrow territory might at first glance lead one to expect. Any single section of this book, and all of them together, clearly exemplify this point. It is essential to recognize the full range of normal, with its anomalies. The same may be said for the range of pancreatitis, carcinoma, and calculus. As one acquires comprehensive experience of the territory, one realizes that as much as clinical history, it is the knowledge of fine-tuned relationships, both contrasts and similarities, among these various expressions of disease that leads to diagnosis.

Overall, the diagnostic accuracy of ERCP is high. For carcinoma, accuracy rates of 92–95% are quoted [Silvis[8], Freeney[1,2]]. Nix[3] devised a complicated formula, based on discriminant analysis, to determine an allocation rule based on eight ERCP criteria (especially changes in the small ducts). This method produced 100% accuracy in diagnosing carcinomas with associated pancreatitis proximal to the point of obstruction. In chronic pancreatitis, the accuracy rate for ERCP is difficult to determine accurately since mild and moderate cases may demonstrate a normal appearance, especially in remission [Robbins[6]]. Ralls[5] claims 93% accuracy, and emphasizes that the most reliable criterion for the diagnosis is the presence of multiple stenoses. He emphasizes the unreliability of utilizing morphologic changes as the only ERCP criteria in differentiating carcinoma from pancreatitis. Also, in the presence of pancreatitis, carcinoma cannot be diagnosed accurately. In Nix's work mentioned previously, the accuracy in chronic pancreatitis was 92%.

## REFERENCES

### History

1. Anacker H: Rontgenanatomie des pankreas. *Fortschr Rontgenstr* 94:1-13, 1961.

2. Bilbao MK, Dotter CT, Lee TG, Katon RM: Complications of endoscopic retrograde cholangiopancreatography (ERCP). A study of 10,000 cases. *Gastroenterology* 70:314-320, 1976.

3. Classen M, Hellwig H, Rosch W: Anatomy of the pancreatic duct; a duodenoscopic—radiological study. *Endoscopy* 5:14-17, 1973.

4. Dawson W, Langman J: An anatomical—radiological study on the pancreatic duct pattern in man. *Anatomical Record* 139:59-68, 1961.

5. Demling L, Classen M (eds): *Endoscopy of the Small Intestine with Retrograde Pancreatocholangiography.* Thieme, Stuttgart: 1973.

6. Doubilet H, Mulholland JH: Intubation of the pancreatic duct in the human. *Proc Soc Exp Biol Med* 76:113-114, 1951.

7. Hirschowitz BI, Curtiss LE, Peters CW, Pollard HM: Demonstration of a new gastroscope, the "fiberscope." *Gastroenterology* 35:50-53, 1958.

8. Kasugai T, Kuno N, Kobayashi S, Hattori K: Endoscopic pancreatocholangiography. I.The normal endoscopic pancreatocholangiogram. *Gastroenterology* 63:217-226, 1972.

 9. Kasugai T, Kuno N, Kizu M, Kobayashi S, Hattori K: Endoscopic pancreatocholangiography. II.The pathological endoscopic pancreatocholangiogram. *Gastroenterology* 63:227-234, 1972.

10. Kreel L, Sandin B, Slavin G: Pancreatic morphology. A combined radiological and pathological study. *Clin Radiol* 24:154-161, 1973.

11. Kreel L, Sandin B: Changes in pancreatic morphology associated with aging. *Gut* 14:962-970, 1973.

12. McCune WS, Shorb PE, Moscovitz H: Endoscopic cannulation of the ampulla of Vater: a preliminary report. *Ann Surg* 167:752-756, 1968.

13. Millbourn E: On the excretory ducts of the pancreas in man with special reference to their relations to each other, to the common bile duct and to the duodenum. A radiological and anatomical study. *Acta Anat* 9:1-34, 1950.

14. Millbourn E: Calibre and appearance of the pancreatic ducts and relevant clinical problems. A roentgenographic and anatomical study. *Acta Chir Scand* 118:286-303, 1959/1960.

15. Oi L, Takemoto T, Kondo T: Fiberduodenoscope: direct observations of the papilla of Vater. *Endoscopy* 3:101-103, 1969.

16. Oi L: Fiberduodenoscopy and endoscopic pancreatocholangiography. *Gastrointest Endosc* 17:59-62, 1970.

17. Rabinov KR, Simon M: Peroral cannulation of the ampulla of Vater for direct cholangiography and pancreatography. Preliminary report of a new method. *Radiology* 85:693-697, 1965.

18. Rienhoff WF, Pickrell KL: Pancreatitis. An anatomic study of the pancreatic and extrapancreatic biliary systems. *Arch Surg* 51:205-219, 1945.

19. Rohrmann CA, Silvis SE, Vennes JA: Evaluation of the endoscopic pancreatogram. *Radiology* 113:297-304, 1974.

20. Sarles H, Sarles JC, Camatte R, Muratore R, Gaini M, Guien C, Pastor J, LeRoy F: Observation on 205 confirmed cases of acute pancreatitis, recurring pancreatitis, and chronic pancreatitis. *Gut* 6:545-559, 1965.

21. Takagi K, Ikeda S, Nakagawa Y, Sakaguchi N, Takahashi T, Kumakura K, Maruyama M, Someya N, Nakano H, Takada T, Takekoshi T, Kin T: Retrograde pancreatography and cholangiography by fiber duodenoscope. *Gastroenterology* 59:445-452, 1970.

22. Takagi K, Ikeda S, Nakagawa Y, Kumakura K, Maruyama M, Someya N, Takada T, Takekoshi T, Kin T: Endoscopic cannulation of the ampulla of Vater. *Endoscopy* 2:107-115, 1970.

23. Waldron RL, Bragg DG, Daly WJ, Seaman WB: Experimental reflux pancreatography and cholangiography. *Radiology* 88:357-360, 1967.

24. Watson WC: Direct vision of the ampulla of Vater through the gastroduodenal fiberscope. *Lancet* 1:902-903, 1966.

## Technique

1. Bilbao MK, Dotter CT, Lee TG, Katon RM: Complications of endoscopic retrograde cholangio-pancreatography (ERCP). A study of 10,000 cases, *Gastroenterology* 70:314-320, 1976.

2. Gregg JA: Detection of bacterial contamination of the pancreatic ducts in patients with pancreatitis and pancreatic cancer during endoscopic cannulation of the pancreatic duct. *Gastroenterology* 73:1005-1007, 1977.

3. Kasugai T, Kuno N, Kizu M: Manometric endoscopic retrograde pancreatocholangiography; technique, significance and evaluation. *Amer J Digest Dis* 19:485-502, 1974.

4. McCune WS, Shorb PE, Moscovitz H: Endoscopic cannulation of the ampulla of Vater: a preliminary report. *Ann Surg* 167:752-756, 1968.

5. Osnes M, Skjennald A, Larsen S: A comparison of a new non-ionic (Metrizamide) and a dissociable (Metrizoate) contrast medium in endoscopic retrograde pancreatography (ERP). *Scand J Gastroenterol* 12:821-825, 1977.

6. Takagi K, Ikeda S, Nakagawa Y, Sakaguchi N, Takahashi T, Kumakura K, Maruyama M, Someya N, Nakano H, Takada T, Takekoshi T, Kin T. Retrograde pancreatography and cholangiography by fiber duodenoscope. *Gastroenterology* 59:445-452, 1970.

7. Takemoto T, Kasugai T, (eds): *Endoscopic Retrograde Cholangiopancreatography.* Igaku-Shoin, New York, 1979, p. 3.

8. Waldron RL: Reflux pancreatography. An evaluation of contrast agents for studying the pancreas. *Am J Roentgenol* 104:632-640, 1968.

## Selection and Choice of Radiologic Procedure

1. Barkin J, Vining D, Miale A, Gottlieb S, Redlhammer DE, Kalser MH: Computerized tomography, diagnostic ultrasound, and radionuclide scanning. Comparison of efficacy in diagnosis of pancreatic carcinoma. *JAMA* 238:2040-2042, 1977.

2. DiMagno EP, Malagelada JR, Taylor WF, Go VLW: A prospective comparison of current diagnostic tests for pancreatic cancer. *N Engl J Med* 297:737-742, 1977.

3. Ferrucci JT, Wittenberg J, Black EB, Kirkpatrick RH, Hall, DA: Computed body tomography in chronic pancreatitis. *Radiology* 130:175-194, 1979.

4. Ferrucci JT, Mueller PR, Harbin WP: Percutaneous transhepatic biliary drainage. Technique, results and applications. *Radiology* 135:1-3, 1980.

5. Foley WD, Stewart ET, Lawson TL, Geenan J, Loguidice J, Maher L, Unger GF: Computed tomography, ultrasonography, and endoscopic retrograde cholangiopancreatography in the diagnosis of pancreatic disease: a comparative study. *Gastrointest Radiol* 5:29-35, 1980.

6. Freeny PC, Ball TJ, Ryan J: Impact of new diagnostic imaging modalities on pancreatic angiography. *Am J Roentgenol* 133:619-624, 1979.

7. Freeny PC, Ball TJ: Endoscopic retrograde cholangiopancreatography (ERCP) and percutaneous transhepatic cholangiography (PTC) in the evaluation of suspected pancreatic carcinoma: diagnostic limitations and contemporary roles. *Cancer* 47:1666-1678, 1981.

8. Gowland M, Warwick F, Kalantzis, N, Braganza J: Relative efficiency and predictive value of ultrasonography and endoscopic retrograde pancreatography in diagnosis of pancreatic disease. *Lancet* 2:190-193, 1981.

9. Hansson JA, Hoevels J, Simert G, Tylen U, Vang J: Clinical aspects of nonsurgical percutaneous transhepatic bile drainage in obstructive lesions of the extrahepatic bile ducts. *Ann Surg* 189:58-61, 1979.

10. Lawson TL. Sensitivity of pancreatic ultrasonography in the detection of pancreatic disease. *Radiology* 128:733-736, 1978.

11. Lukes PJ, Rolny P, Nilson AE, Gamklou R: Hypotonic duodenography and endoscopic retrograde pancreatography in the diagnosis of pancreatic disease. *Acta Radiol (Diagn)* 22:145-150, 1981.

12. Moss AA, Federle M, Shapiro HA, Ohto M, Goldberg H, Korobkin M, Clemett A: The combined use of computed tomography and endoscopic retrograde cholangiopancreatography in the assessment of suspected pancreatic neoplasm: a blind clinical evaluation. *Radiology* 134: 159-163, 1980.

13. Ralls PW, Halls J, Renner I, Juttner H: Endoscopic retrograde cholangiopancreatography (ERCP) in pancreatic disease. A reassessment of the specificity of ductal abnormalities in differentiating benign from malignant disease. *Radiology* 134:347-352, 1980.

14. Ring EJ, Oleaga JA, Freiman DB, Husted JW, Lunderquist A: Therapeutic applications of catheter cholangiography. *Radiology* 128:333-338, 1978.

15. Suzuki T, Manabe T, Tani T, Tobe T: Angiography and pancreatoductography in resectable carcinoma of the pancreas. *Acta Radiol (Diagn)* 21:587-591, 1980.

16. Wittenberg J, Mueller PR, Ferrucci JT, Simeone JF, van Sonnenberg E, Neff CC, Palermo RA, Isler RJ: Percutaneous biopsy of abdominal tumors using 22-gauge needles: further observations. *Am J Roentgenol* 139:75-80, 1982.

17. Yamanaka T, Kimura K: Differential diagnosis of pancreatic mass lesion with percutaneous fine-needle aspiration biopsy under ultrasonic guidance. *Dig Dis and Sci* 24:694-699, 1979.

## Radiologic Interpretation

1. Freeny PC, Bilbao MK, Katon RM: "Blind" evaluation of endoscopic retrograde cholangiopancreatography (ERCP) in the diagnosis of pancreatic carcinoma: the "double duct" and other signs. *Radiology* 119:271-274, 1976.

2. Freeny PC, Ball TJ: Evaluation of endoscopic retrograde cholangiopancreatography and angiography in the diagnosis of pancreatic carcinoma. *Am J Roentgenol* 130:683-691, 1978.

3. Nix GAJJ, Schmitz PIM: Diagnostic features of chronic pancreatitis distal to benign and to malignant pancreatic duct obstruction. *Diagnostic Imaging* 50:130-137, 1981.

4. Ohto M, Ono T, Tsuchiya Y, Saisho, H (eds): *Cholangiography and Pancreatography.* University Park Press, Baltimore, 1978.

5. Ralls PW, Halls J, Renner I, Juttner H: Endoscopic retrograde cholangiopancreatography (ERCP) in pancreatic disease. A reassessment of the specificity of ductal abnormalities in differentiating benign from malignant disease. *Radiology* 134:347-352, 1980.

6. Robbins AH, Messian RA, Widrich WC, Paul RE Jr, Norton RA, Schimmel EM, Ogoshi K: Endoscopic pancreatography: an analysis of the radiologic findings in pancreatitis. *Radiology* 113:293-296, 1974.

7. Rohrmann CA, Silvis SE, Vennes JA: The significance of pancreatic ductal obstruction in differential diagnosis of the abnormal endoscopic retrograde pancreatogram. *Radiology* 121:311-314, 1976.

8. Silvis SE, Rohrmann CA, Vennes JA. Diagnostic accuracy of endoscopic retrograde cholangiopancreatography in hepatic, biliary, and pancreatic malignancy. *Ann Int Med* 84:438-440, 1976.

9. Stewart ET, Vennes JA, Geenen JE (eds): *Atlas of Endoscopic Retrograde Cholangiopancreatography.* Mosby, Saint Louis, 1977.

10. Takemoto T, Kasugai T (eds): *Endoscopic Retrograde Cholangiopancreatography.* Igaku-Shoin, New York, 1979.

## BIBLIOGRAPHY

Ariyama J: *Radiology in Disorders of the Liver, Biliary Tract and Pancreas, with Special Reference to the Use of Angiography in the Diagnosis of Resectable Tumors.* Igaku-Shoin, New York, 1981.

Berk RN, Clemett AR: *Radiology of the Gallbladder and Bile Ducts.* Saunders, Philadelphia, 1977.

Bockus HL(ed): *Gastroenterology,* 3rd ed. Saunders, Philadelphia, 1976.

Demling L, Classen M (eds): *Endoscopy of the Small Intestine with Retrograde Pancreatocholangiography.* Thieme, Stuttgart, 1973.

Eaton SB, Ferrucci JT: *Radiology of the Pancreas and Duodenum.* Saunders, Philadelphia, 1973.

Hatfield PM, Wise RE: *Radiology of the Gallbladder and Bile Ducts.* Williams & Wilkins, Baltimore, 1976.

Howat HT, Sarles H (eds): *The Exocrine Pancreas.* Saunders, London, 1980.

Margulis AR, Burhenne HJ (eds): *Alimentary Tract Radiology.* 2nd ed. Mosby, Saint Louis, 1973.

McNulty JG: *Radiology of the Liver.* Saunders, Philadelphia, 1977.

Meyers MA: *Dynamic Radiology of the Abdomen. Normal and Pathologic Anatomy.* Springer-Verlag, New York, 1976.

Ohto M, Ono T, Tsuchiya Y, Saisho H (eds): *Cholangiography and Pancreatography.* University Park Press, Baltimore, 1978.

Pernkopf E (ed): *Atlas of Topographical and Applied Human Anatomy.* Saunders, Philadelphia, 1964.

Stewart ET, Vennes JA, Geenen JE (eds): *Atlas of Endoscopic Retrograde Cholangiopancreatography.* Mosby, Saint Louis, 1977.

Takemoto T, Kasugai T (eds): *Endoscopic Retrograde Cholangiopancreatography.* Igaku-Shoin, New York, 1979.

Whalen JP: *Radiology of the Abdomen. Anatomic Basis.* Lea & Febiger, Philadelphia, 1976.

Williams PL, Warwick R (eds): *Gray's Anatomy.* 36th ed. Saunders, Philadelphia, 1980.

Wright R, Alberti KGMM, Karran S, Millward-Sadler GH (eds): *Liver and Biliary Disease. Pathophysiology, Diagnosis, Management.* Saunders, Philadelphia, 1979.

# II

# Normal ERCP

# INTRODUCTION

This section of normal ERCP presents the full range of cases typically seen in clinical practice, particularly with respect to the clinical circumstances which call for ERCP. While the clinical histories encompass a wide range of symptoms, these must not be confused with indications for ERCP which are, as noted in the Introductory Chapter, remarkably few. Consistent with this construct, Cases 1, 3, 6, and 7 were performed for undiagnosed abdominal pain, Case 4 for jaundice with the primary focus on excluding extrahepatic bile duct obstruction, and Cases 2 and 8 to seek a primary lesion in both confirmed (Case 2) and suspected (Case 8) metastatic carcinoma.

Although many of these patients in fact had disease of other systems, ERCP demonstrates throughout Section II *normal* pancreatic and biliary anatomy, where anomaly is the single significant variant. Radiologically, anatomy as seen on a radiograph is viewed as normal or abnormal because an internal, abstract perception of normal exists, rather like a visual template, to which the individual case may be compared. The cases in this section thus serve to familiarize the reader with the comprehensive range of normal pancreatic and biliary anatomy. The normal images on the radiographs which follow are evoked during the diagnostic process and are, therefore, essential to interpretation. In this regard, adherence to a strict millimeter range as the standard for normal duct caliber is not useful as a substitute for visual familiarity with the full range of normal radiologic anatomy seen in clinical practice.

The accuracy of interpretation of ERCP is wholly dependent on quality radiography, and specifically on meticulous radiographic technique. In this section, as throughout, this aspect of ERCP is emphasized on nearly a case-by-case basis in the "Comment" to the cases.

In sum, features to note for Section II include: a) typicality of radiographic anatomy presented; b) narrow range of true indications for ERCP; c) normal radiographic anatomy as a foundation for visual template formation; and d) the importance of technique.

## CASE 1
### (Accompanying Radiographs A & B)

**Clinical History**

A 38-year-old white male presented with a history of alcoholism and polycystic renal disease associated with recent onset of abdominal pain. ERCP was performed to rule out associated cystic disease of the pancreas.

**Radiographic Findings**

A. Intravenous pyelogram shows changes of bilateral polycystic disease. Note renal enlargement and deformity of the collecting systems which are splayed, distorted, and show spidery elongation, especially in the lower poles.

B. ERCP in the AP projection shows the endoscope passing along the greater curvature of the stomach, and the endoscopy catheter (arrowheads) approximately 5 mm within the pancreatic duct. There is filling of the entire pancreatic duct which courses from the level of the midbody of L-1 vertebra so that the tail is more cephalad at the level of mid-T-12. There is progressive smooth tapering from an initial caliber of 2 mm in the head to under 1 mm in the tail.

**Diagnosis**

Normal pancreatic duct.

**Comment**

The normal pancreas is usually oriented vertically to a greater or lesser extent, with the tail more cephalad than the head as in this case; less often, it has a transverse orientation. The pancreatic duct, thus, reflects the contour and orientation of the pancreas itself. The cephalic and anterior curve seen in the head and distal body in this case reflects the pancreas passing over the vertebral body. Note that injection of the opaque medium in this projection results in visualization primarily of the middle and distal thirds of the duct (body and head portions). This projection also accounts for the foreshortened appearance of the pancreatic duct in the head. Because of these limitations oblique views are standard. (See Case 2.)

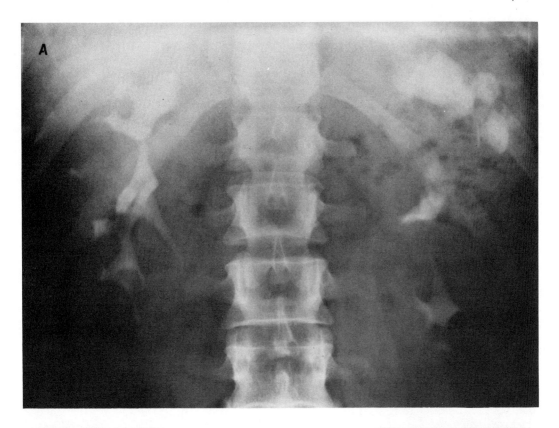

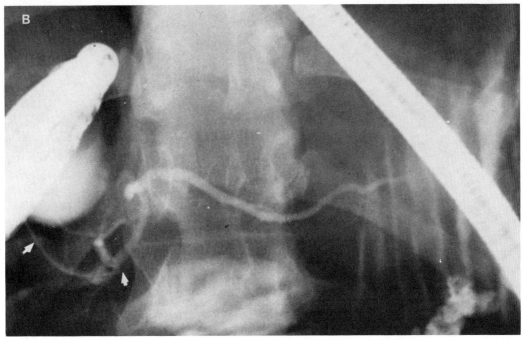

Case 1, Radiographs A & B

**CASE 2**
(Accompanying Radiographs A & B)

### Clinical History

A 54-year-old white male presented with metastatic carcinoma of the left ischium, weight loss, and a four-month history of pain on flexing left hip. Alkaline phosphatase and other liver function tests (LFT) were normal. ERCP was performed in search of a primary site of carcinoma.

### Radiographic Findings

A. The shallow RPO projection shows a normal pancreatic duct entirely visualized with moderate overfilling and extensive acinar filling (1). The ampulla is seen at the level of L-2 (arrow) and the tail at T-12/L-1 interspace.

B. The LPO view shows filling of the pancreatic duct as far as the tail (proximal third) with extensive side-branch filling throughout (arrows). The caliber of the pancreatic duct ranges from 4 mm in the head to 2 mm in the tail.

### Diagnosis

Normal pancreatic duct.

### Comment

In contrast to Case 1, this pancreas is much more vertical in orientation and ascends nearly two vertebral bodies from head to tail which represents a typical appearance. The caliber of this pancreatic duct is at the upper range of normal which is the consequence of the pressure used during injection. Thus, the true size of the duct, measured in a nondistended state, is approximately one-half of the caliber seen here. Because the pancreatic duct is distensible, overfilling will distort accurate measurement. Usually injection is monitored fluoroscopically and ceases when any evidence of side-branch filling is noted. This usually occurs prior to acinarization. However, if one wishes to verify whether nonfilling of a segment is an artifact of technique, slight overfilling, up to and including side branches and acini, may be necessary. Deliberate overfilling should be used, however, only when other techniques fail to produce complete filling of the pancreatic duct. The problem with filling to the degree seen here is the potential for an increased incidence of pancreatitis. With acinarization there is passage of the contrast material from the small branches of the pancreatic duct into the acini of the pancreatic parenchyma providing the anatomic basis for an inflammatory response. Note that acinarization with overfilling, as in this case, usually predominates in one region only and, although this is typical, the phenomenon is not well understood. In this case the filling defect on Radiograph B in the distal pancreatic duct is artifact, due to an air bubble, and overfilling resulted from attempts to dispel it.

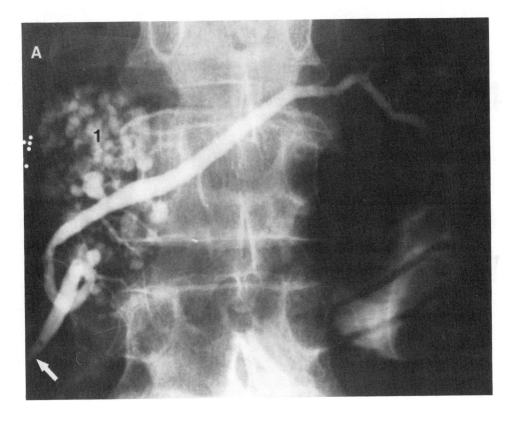

Case 2, Radiograph A

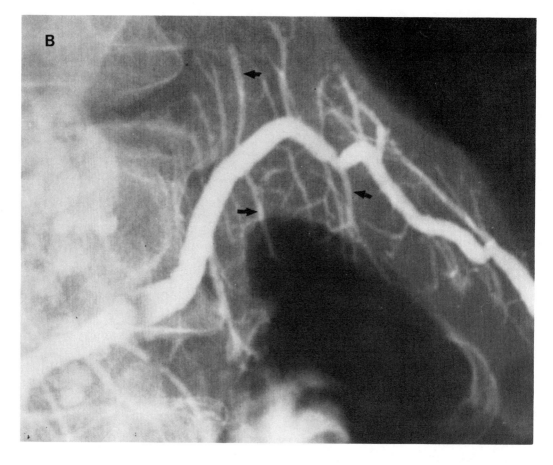

Case 2, Radiograph B

# NOTES

## CASE 3
### (Accompanying Radiographs A-C)

### Clinical History

A 60-year-old black male with chronic alcoholism presented with a 6-week history of abdominal pain of undetermined origin. On examination there was a clinically palpable tender midepigastric mass, he was febrile, and bilirubin and urine amylase were slightly elevated.

### Radiographic Findings

A. Gallium scan shows increased activity over the upper central abdomen in the region of the head of the pancreas (arrows), indicating the presence of an inflammatory process.

B. ERCP, performed to confirm Gallium scan findings, shows a normal pancreatic duct. The duodenal bulb (arrows) and stomach (1) are visualized and overlie the distal pancreatic and common bile ducts in this RPO projection.

C. The RPO view also demonstrates the pancreas oriented obliquely with the tail (1) about two vertebral bodies cephalad to the ampulla (2). Note the smooth, progressive tapering from head to tail which is normal, as is the slight degree of filling of side branches in the body and head.

### Diagnosis

Normal pancreatic duct.

### Comment

During injection of the pancreatic duct it is normal for some contrast material to spill around the catheter into the duodenal lumen and pass along into the jejunum. (See Case 5, Radiograph A.) It is also normal for the material to be carried proximally, by reverse peristalsis, to the duodenal bulb and stomach. Note that the duodenal bulb is closely related to the lateral aspect of the distal common bile duct, and that the body of the pancreatic duct is intimately related to the posteroinferior surface of the stomach. Duodenal diverticula opacify such that they may be confused with bulb, but may be distinguished by identification of the site of origin of filling. For example, those diverticula which may occur at this site arise from the second and first portions of the duodenum. Duodenal bulb is always visualized at the first portion (by definition), has a slightly triangular configuration with greater or lesser fornices and apex, and has a pyloric canal entering its base. (See Case 8, Radiograph A.) Gallbladder also may be mistaken for bulb, although usually identification of the cystic duct avoids confusion.

Note that a small abscess was found in the head of the pancreas at surgery. Such false-negative findings may be expected with ERCP where the inflammatory lesion is less than 1 cm in diameter (as was the case here). Because deformity of the pancreatic duct is a consequence of parenchymal changes, pathophysiologic processes must either involve so substantial a portion of pancreatic tissue, or be in such close proximity to the pancreatic duct, as to effect alterations of the duct itself.

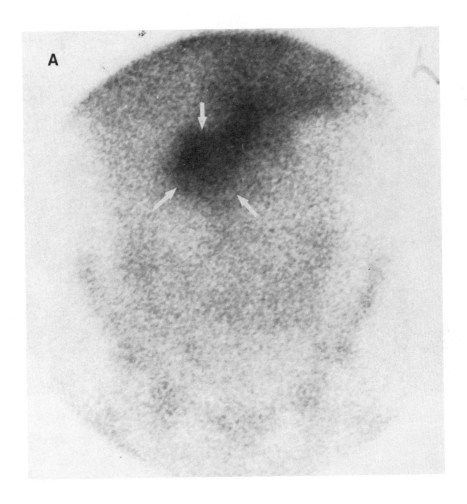

Case 3, Radiograph A

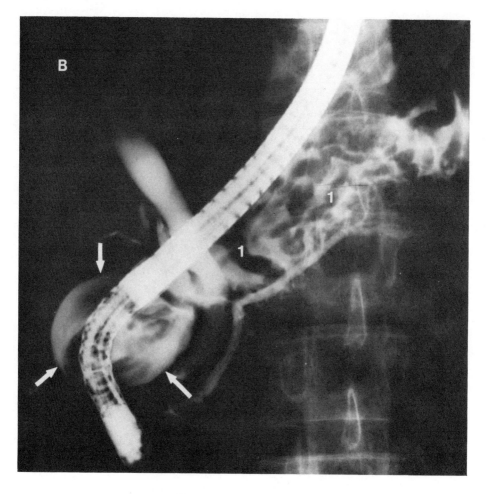

**Case 3, Radiograph B**

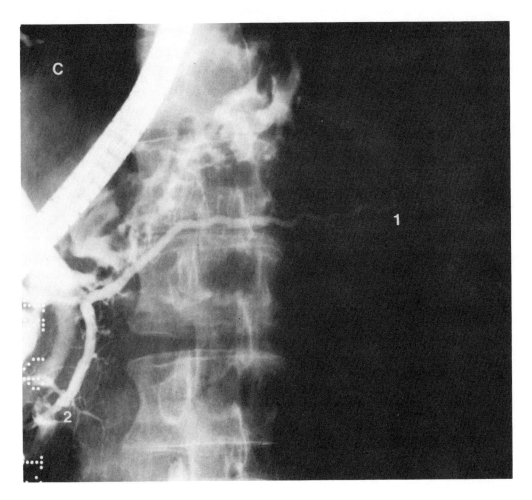

Case 3, Radiograph C

## CASE 4
(Accompanying Radiographs A & B)

### Clinical History

A 54-year-old white female presented with a history of chronic persistent hepatitis of 7 years duration. ERCP was performed to rule out extrahepatic obstruction since LFT were inconsistent with this diagnosis.

### Radiographic Findings

A. Normal pancreatic duct visualized in its entirety (shallow RPO). The slight angulation (x) seen here at the junction of the body and head is not uncommon. There is very slight roughening of the margin.

B. The distal common bile duct (1) and pancreatic duct (2) lie in close proximity as they traverse the head of the pancreas (AP). The endoscope causes a slight extrinsic pressure deformity (arrows) on the common bile duct as it curves through the pylorus.

### Diagnosis

Normal pancreatic duct.

### Comment

The normal pancreatic duct shows a relatively smooth margin ranging from entirely regular to slightly roughened as in this case. Distinct scalloping and caliber variation are not normal and usually indicate the presence of a pathological process, such as chronic pancreatitis. Note that the distal portions of the pancreatic duct and common bile duct are in parallel, adjacent relationship. This is particularly significant pathophysiologically since their proximity accounts for their simultaneous involvement by disease processes in the head.

Generally, oblique views are a necessary complement to AP projections: a) to compensate for the AP foreshortening of the head; b) to visualize the tail and ampulla; and c) to eliminate other organ or instrument interference such as the extrinsic pressure deformity produced here by the endoscope.

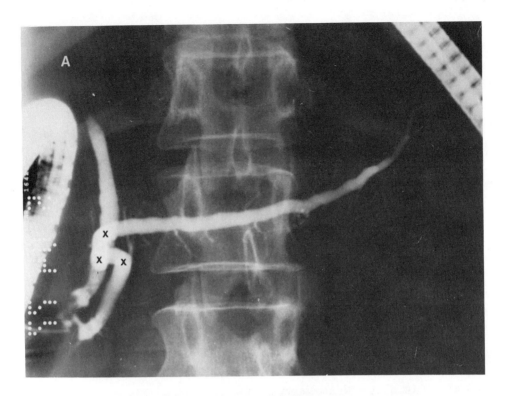

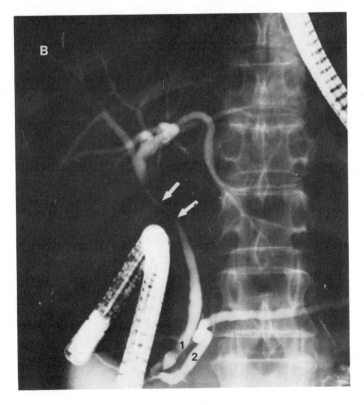

Case 4, Radiographs A & B

## CASE 5
### (Accompanying Radiograph A)

### Clinical History

A 60-year-old black male presented with a 6-month history of epigastric pain, the etiology of which could not be established in spite of complete evaluation.

### Radiographic Findings

A. ERCP in the RPO view demonstrates a normal pancreatic duct as well as an accessory duct (arrowheads). Note the filling of the distal duodenum and proximal jejunum (b).

### Diagnosis

Normal pancreatic duct with accessory duct.

### Comment

This case presents an excellent example of an accessory duct (arrowheads) of about 1 mm diameter. It enters the main pancreatic duct several centimeters proximal to the ampulla (a). Accessory ducts may vary in caliber (1–2 mm) and in length (3–5 cm). The curvature of the pancreatic duct at the accessory duct may be marked, as in Case 4, Radiograph A, x, although this is not invariably the case.

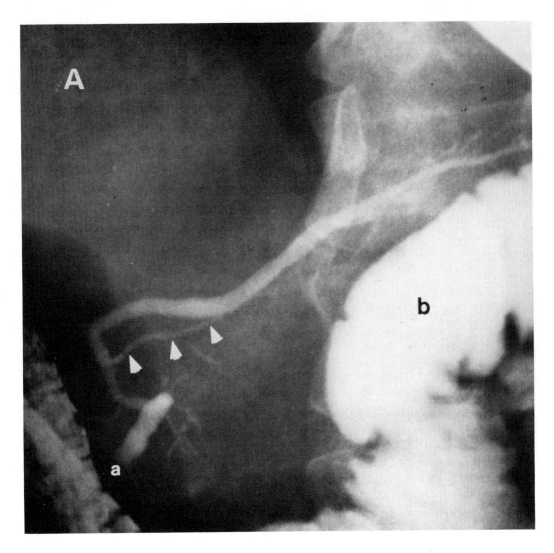

Case 5, Radiograph A

## CASE 6
### (Accompanying Radiographs A & B)

### Clinical History

A 28-year-old white male presented with a 3-week history of severe epigastric pain. Relevant history included malignant hypertension with nephrosclerosis, cadaver kidney transplant, and dialysis. On current admission, upper GI series demonstrated C-loop changes suggesting pancreatitis. Serum calcium and alkaline phosphatase were slightly elevated. ERCP was performed to rule out pancreatitis associated with chronic renal disease and dialysis.

### Radiographic Findings

A & B. ERCP shows a normal pancreatic duct (pd), with an accessory duct approximately 2 mm in diameter (arrowheads), from which moderate side-branch filling occurs.

### Diagnosis

Normal pancreatic duct with accessory duct.

### Comment

The caliber of the accessory duct here is greater than that seen in Case 5, and represents the upper limit of the normal range. With accessory ducts of this size, potential for confusion with a partially filled pancreatic duct exists. In order to distinguish these two structures: a) note the presence of a segment of pancreatic duct distal to the accessory duct such that usually the accessory duct enters the pancreatic duct proximal to the ampulla; (b) where the preceding is not the case, and the accessory duct opens directly into the ampulla (or, more rarely, into the duodenum), careful search must be made for the orifice of the pancreatic duct.

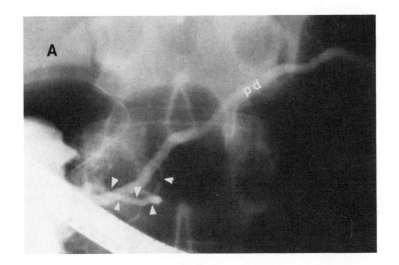

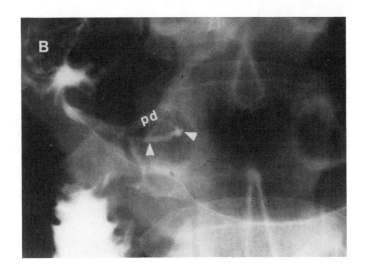

Case 6, Radiographs A & B

## CASE 7
### (Accompanying Radiographs A & B)

### Clinical History

A 45-year-old white male presented with a history of chronic right upper quadrant pain. Physical examination was negative and LFT were normal. ERCP was performed to rule out pancreatitis.

### Radiographic Findings

A ERCP demonstrates a slender normal pancreatic duct (1), common bile duct (2),
& filling of the intrahepatic branches (3), cystic duct (4), and partial filling of the
B. gallbladder (5).

### Diagnosis

Normal pancreatic duct and biliary system.

### Comment

Although not routine, cannulation of the common bile duct with injection of contrast material and visualization of the biliary tree is necessary when the clinical history is equivocal and prior cannulation of the pancreatic duct shows it to be normal. Common duct cannulation should demonstrate the common duct in its entirety and show a variable degree of intrahepatic bile duct filling. The extent of system visualization necessary is determined by clinical requirements. Cystic duct visualization is regularly achieved, but gallbladder is less often opacified (about 80%) and, in some cases, defies deliberate attempts to do so.

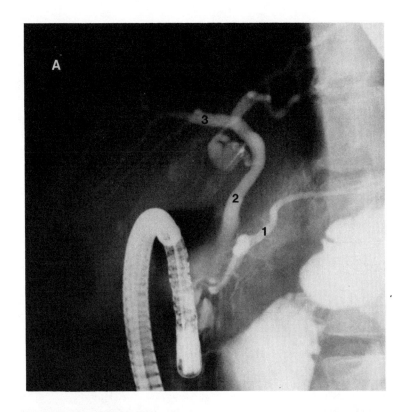

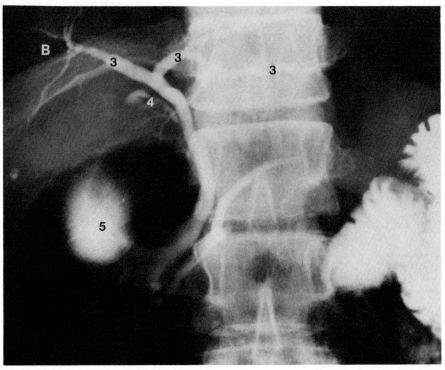

Case 7, Radiographs A & B

## CASE 8
(Accompanying Radiographs A-D)

**Clinical History**

A 55-year-old white male presented with a 60-pound weight loss sustained over the previous year, was otherwise asymptomatic, and physical examination on the current admission was essentially negative.

**Radiographic Findings**

A   The duodenal bulb (1) mimics a partially filled gallbladder. Note filling of the
&   cystic duct (2), as well as of the accessory pancreatic duct (3). The terminal
B.  portions of the pancreatic duct and common bile duct in the ampullary segment (4) are slightly narrowed.

C.  With further injection, there is filling of the gallbladder infundibulum (5) which is identified by the presence of the valves of Heister (arrowheads).

D.  The endoscopy catheter (arrowheads) is now located in the distal common bile duct.

**Diagnosis**

Normal pancreatic duct and biliary system.

**Comment**

Clinically, the weight loss was determined ultimately to be associated with schizophrenia. Together, Cases 7 and 8 present a comprehensive illustration of a complete ERCP demonstrating most of the features of normal anatomy. Because the duodenal bulb often is opacified by retrograde peristalsis during the procedure, it may mimic the gallbladder. Identification of the linear filling defects of the valves of Heister is a useful differential feature for the gallbladder neck and cystic duct. Note that there is less filling of the distal common bile and pancreatic ducts seen in Radiograph B than in A due to contraction of the ampullary sphincter in B, which also accounts for the slight narrowing at the ampulla seen in Radiograph B. With complete contraction of the sphincter, the lumen is obliterated and no filling will be appreciated. Thus, precise timing of both cannulation and filming is essential, and not only represent good technique that permits appropriate filling and visualization, but demonstrate as well that such technique provides the basis of diagnostic capability since it assures the absence of technical or operator error when pathologically related phenomena, such as stenosis or obstruction, visualize.

The assessment of any procedure as complete is a clinical judgment overall. However, minimally a complete procedure must include visualization of the entire pancreatic duct, and nonfilling of any segment represents a significant finding which must be accounted for. Further, one may optionally elect to enhance the scope of the procedure to include common bile duct cannulation with sequential visualization of the a) common bile duct itself; b) common bile duct and intra-hepatic branches; and c) common bile duct, intrahepatic branches, and gallbladder. Application of all of these possibilities will be presented in succeeding sections of this book where the use or emphasis of each in relation to a specific pathologic abnormality together with the clinical configuration will be demonstrated.

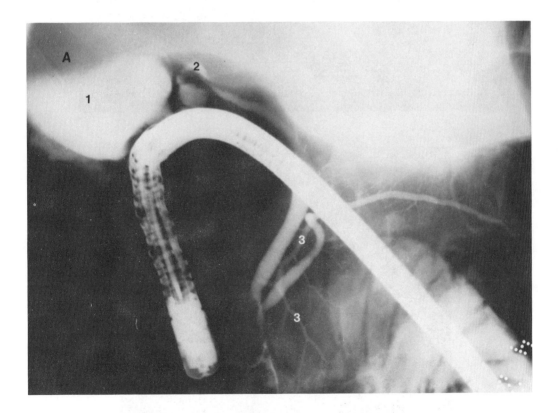

Case 8, Radiograph A

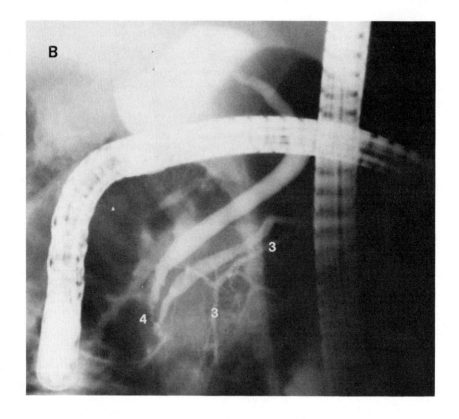

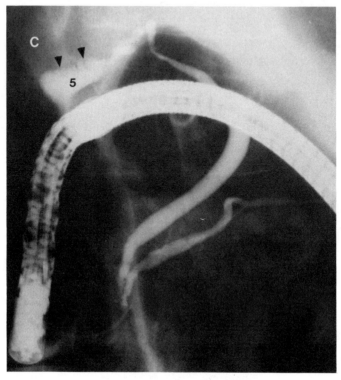

Case 8, Radiographs B & C

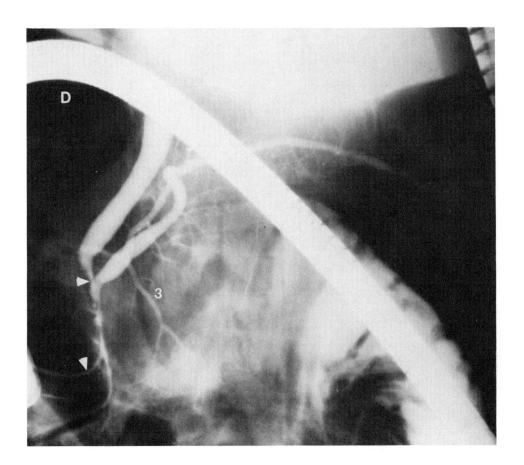

Case 8, Radiograph D

# III

# Calculi

# INTRODUCTION

ERCP is the procedure of choice for diagnosis of calculi in the pancreatic duct itself. It is a secondary procedure for calculi in the biliary system and gallbladder where ultrasound, oral cholecystography, and intravenous cholangiography are the mainstays of diagnosis.

Indications for ERCP include jaundice, abdominal pain (rarely, establishment of the cause of known pancreatitis), and diagnostic failure of primary procedures, especially in the case of suspected calculus.

In the case of the pancreatic duct ERCP is used particularly for its ability over other modalities to visualize very small calculi as well as associated conditions such as stricture or duct ectasia. Ultrasound, although unable to show the smallest calculi (under 2 mm diameter) is easily the procedure of choice for demonstrating calculi in the common bile duct and gallbladder. It also has the particular advantage of not being limited in application by the presence of jaundice, as are oral cholecystography and intravenous cholangiography.

For the cases presented, the focus is on tailoring the diagnostic "package" to the needs of the individual clinical circumstance. Often, multiple procedures may need to be employed in a series before the diagnosis emerges.

For visual template formation note that the hallmark of calculus, radiographically, is the presence of a filling defect which is sharply defined, facetted, and surrounded by contrast material.

**CASE 9**
(Accompanying Radiographs A-D)

## Clinical History

A 74-year-old white female presented with a history of abdominal pain, mildly abnormal LFT for 3 months prior to current admission, and weight loss for 6 months. There was nonvisualization on two attempts at oral cholecystography, and ERCP was then performed.

## Radiographic Findings

A. Intravenous cholangiogram (IVC) shows a single calculus (arrowheads) in a markedly dilated distal common bile duct (1).

B. Retrograde cholangiogram (ERC) demonstrates that several faceted calculi
& visualize (arrowheads) in the distal common bile duct and that these are mobile
C. by virtue of their change in position from Radiograph B to C. Intrahepatic branches (2) are moderately dilated but show no calculi.

D. On retrograde filling, multiple calculi are also demonstrated within the gall-bladder (arrowheads).

## Diagnosis

Calculi, common bile duct, and gallbladder (confirmed at surgery).

## Comment

Here IVC and ERCP were used in conjunction to reach a clinical diagnosis and, generally, ultrasound is considered as well. The procedure of choice in a given case, or the sequence in which a series of studies are employed, is a function of both yield (diagnosis) and clinical imperative. IVC visualization, as in this case, of a single calculus in the common bile duct may or may not be regarded as sufficient clinical information since, for example, the presence of intrahepatic calculi would affect the surgical approach. Conversely, in suitable circumstances, the substantiation of calculi exclusively in the common bile duct permits the consideration of basket retrieval. In this regard, demonstration of impaction versus mobility is essential. Obviously, specialty diagnostic preference with regard to any one or a combination of diagnostic studies must be seen as secondary to the unique clinical setting of each patient and evaluations should thus be tailored accordingly. The significant advantage of ERCP or the modified ERC over such examinations as IVC is that either may be used without the limitations placed by the presence of jaundice as it reflects impaired liver function.

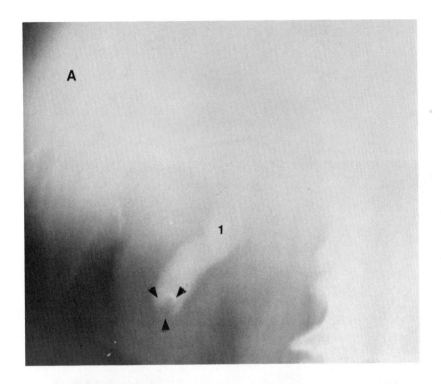

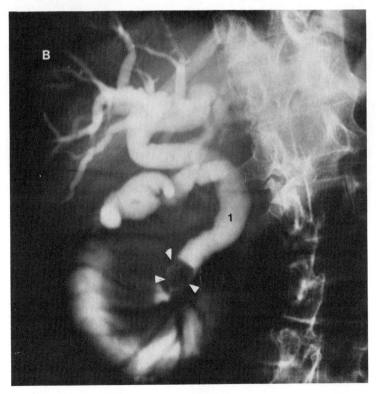

Case 9, Radiographs A & B

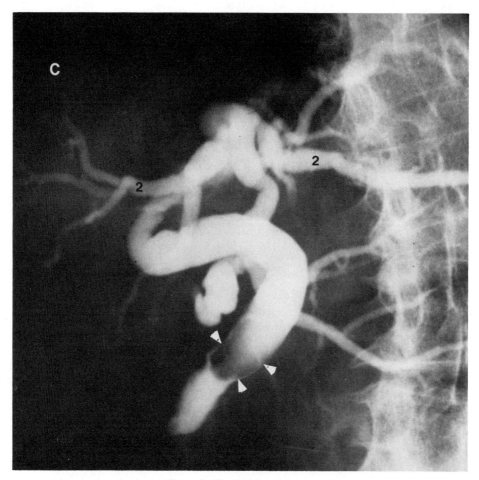

Case 9, Radiograph C

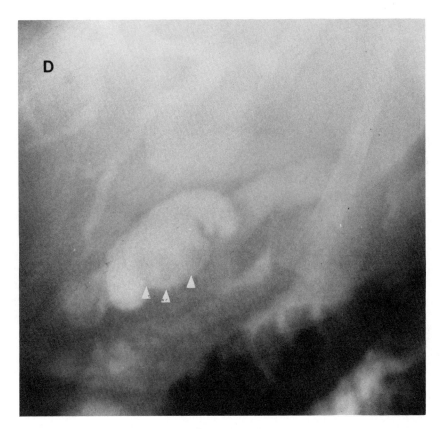

Case 9, Radiograph D

# CASE 10
## (Accompanying Radiograph A)

**Clinical History**

A 79-year-old white male presented with a 2-month history of abdominal pain, distention, anorexia, pruritus, and associated 30-pound weight loss. On examination he was jaundiced (bilirubin, 6 mg) and had hepatomegaly.

**Radiographic Findings**

**A.** Retrograde cholangiogram (ERC) shows two large calculi (arrowheads) in the distal common bile duct.

**Diagnosis**

Common bile duct calculi (confirmed at surgery).

**Comment**

In elderly patients with jaundice and symptoms strongly suggesting carcinoma of the pancreas, ERCP is particularly useful where alternatives, such as laparotomy, are high risk. Because of the potential of continuing impairment of liver function with cases such as this, the rapid definitive diagnosis of the cause of jaundice is especially important to clinical management. Patients tolerate ERCP well provided they are capable of the minimal cooperation required.

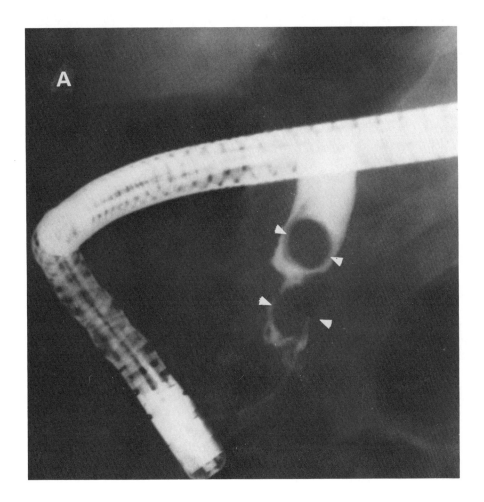

Case 10, Radiograph A

# CASE 11
## (Accompanying Radiographs A-D)

## Clinical History

A 73-year-old black male presented with a recent 10-pound weight loss and an episode of nausea, vomiting, and right lower quadrant pain 2 days prior to the current admission. On examination he was febrile and serum amylase was moderately elevated. Oral cholecystography did not visualize the gallbladder.

## Radiographic Findings

A. Upper GI series shows C-loop mucosal edema and spiking (arrows) suggesting acute pancreatitis.

B. ERCP shows normal pancreatic and common bile ducts.
&
C.

## Diagnosis

Normal pancreatic duct and common bile duct.

D. Repeat oral cholecystogram 2 days after ERCP shows three large calculi in the gallbladder. (Numbers 1, 2, and 3 denote the center of each stone and the gallbladder margin is defined by the dotted line.)

## Comment

With normal findings, the question this case raises concerns that degree of comprehensiveness of an ERCP which may be regarded as complete. The mucosal spiking seen on upper GI series was the significant misleading feature here since it directed attention to pancreatitis and, thus, the pancreatic duct. This, in turn, led to less emphasis on visualization of the entire biliary system (including gallbladder). The judgment here was that complete pancreatic duct and common bile duct visualization provided maximum diagnostic yield possible in view of the likelihood of acute pancreatitis and the potential, thus, for complications were complete filling deliberately pursued. Even in retrospect, and given the oral cholecystogram findings, the C-loop changes may not be attributed definitively to cholecystitis. In summary, the results of the conservative approach may not be gratifying in a given instance when the occasional false-negative occurs. However, this must be balanced against the reduced incidence of complication overall which such an attitude ensures, and particularly with regard to those associated with acinarization.

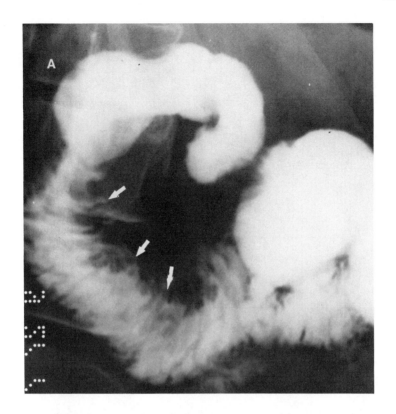

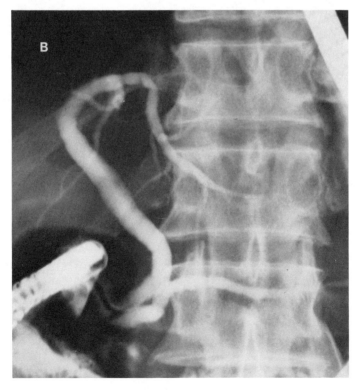

Case 11, Radiographs A & B

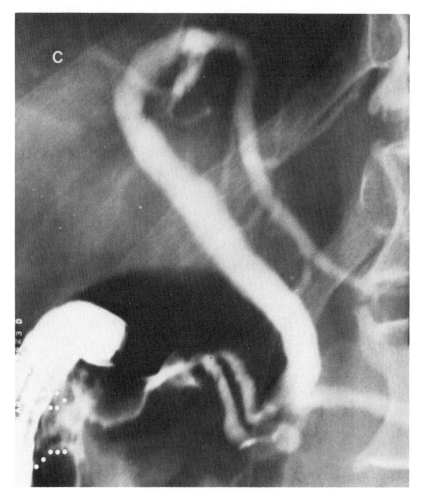

Case 11, Radiograph C

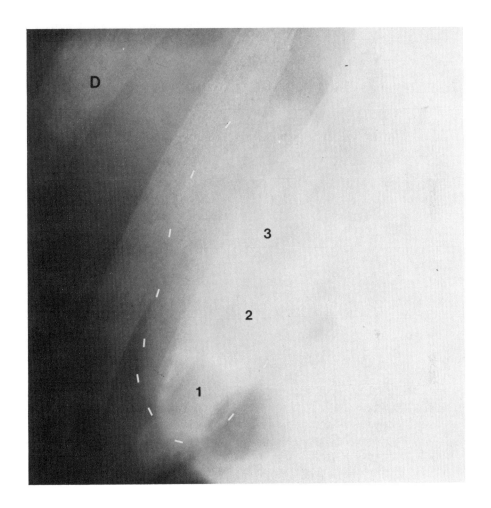

Case 11, Radiograph D

## CASE 12
### (Accompanying Radiographs A-D)

### Clinical History

A 30-year-old white male presented with a 1-year history of intermittent abdominal pain. The serum amylase was elevated on examination and other laboratory studies were normal.

### Radiographic Findings

A. ERCP shows an entirely normal pancreatic duct. Punctate densities (1) are due
& to residual contrast material in the duodenal bulb.
B.

C  With retrograde filling of the cystic duct (2) there is visualization of multiple
&  radiolucent calculi in the partially filled gallbladder (3) best seen in the decu-
D. bitus view (Radiograph D), where the calculi layer is under the influence of gravity (arrowheads).

### Diagnosis

Cholelithiasis.

### Comment

Maneuvering the patient into various positions is an integral part of ERCP examination to obtain adequate visualization of all portions of the pancreatic duct and biliary system. Those appropriate for obtaining gallbladder filling include the right lateral decubitus and right anterior oblique. The former was used in this case to ensure filling, and, more specifically, to achieve the layering of radiolucent calculi above the heavier opacified bile in the fundus of the gallbladder. Thus, in Radiograph D, a layer of small stones may be seen, each of which by itself is virtually undetectable. RAO would be used during the filling phase to complement the decubitus when the former position does not produce adequate filling.

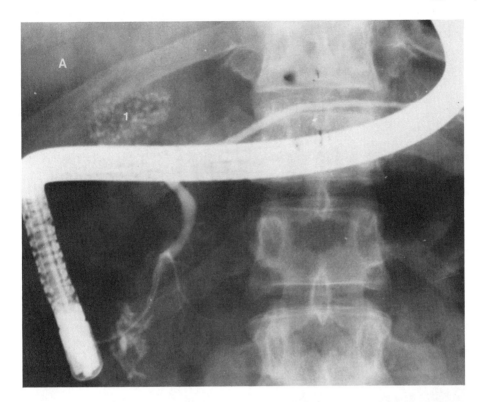

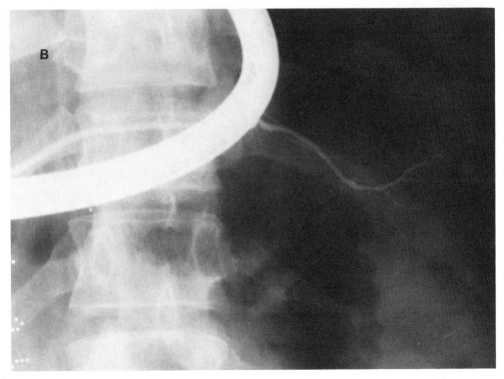

Case 12, Radiographs A & B

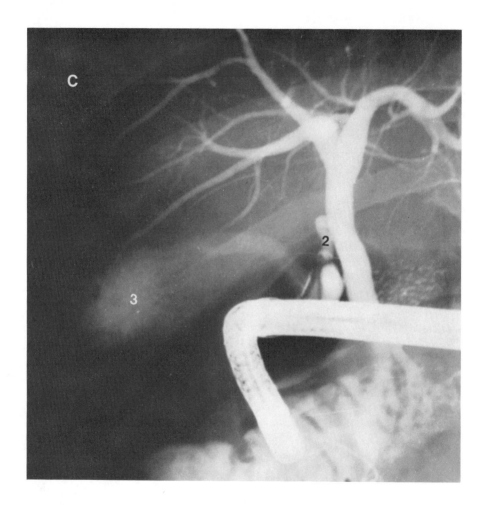

**Case 12, Radiograph C**

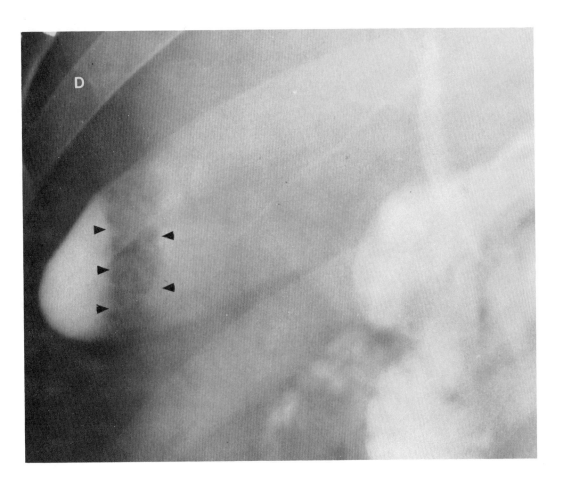

Case 12, Radiograph D

## CASE 13
### (Accompanying Radiographs A & B)

### Clinical History

A 37-year-old white female presented with a 6-year history of Crohn's disease for which multiple bowel resections had been performed. Two years prior to the current admission LFT's were abnormal and attributed to liver disease and short bowel syndrome. One year before the current admission, Graves' disease was also diagnosed. She presented on this occasion with abdominal pain associated with a single episode of passing dark urine. Bilirubin on admission was 1.2 mg.

### Radiographic Findings

A. Retrograde cholangiogram (ERC) shows a radiolucent calculus (arrow) in the distal common bile duct (1).

B. Multiple radiolucent calculi are demonstrated in the gallbladder.

### Diagnosis

Calculi, common bile duct and gallbladder.

### Comment

In contrast to Case 12, the multiple calculi in the gallbladder here are much more clearly appreciated despite their radiolucent nature. This is due to their slightly larger size and more intense opacification of the gallbladder. Decubitus view also was taken, but is not essential to diagnosis here. Note that the solitary calculus in the distal common bile duct accounts for the patient's current symptoms. Presumably, a stone of the size seen here would pass spontaneously, resulting in remission of symptoms, but only until the next stone dislodged from the gallbladder. This case would appear to support the hypothesis that there is a causal relationship between Crohn's disease and the formation of cholesterol calculi in the gallbladder due to deficient absorption of bile salts and resulting distortion of the ratio of bile salts to cholesterol[1].

### Reference

1. Bockus HL: Chronic inflammatory diseases of the intestines (Chap. 74). In *Gastroenterology*, H.L. Bockus (ed.) Saunders, Philadelphia, 1976, p. 583.

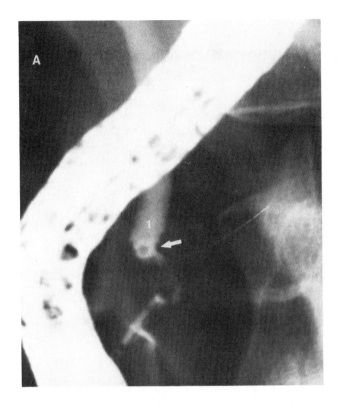

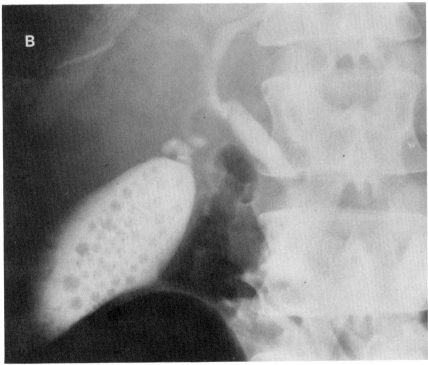

Case 13, Radiographs A & B

## CASE 14
(Accompanying Radiographs A & B)

### Clinical History

An 81-year-old white male presented with a 4-week history of fever, chills, and 10-pound weight loss. He became jaundiced on admission, although physical examination was negative otherwise and upper GI series was normal. The patient stated that he had been told he had a gallstone 5 years previously.

### Radiographic Findings

A & B. Retrograde cholangiogram (ERC) shows severe narrowing of the proximal common bile duct for a length of about 3 cm (arrowheads) due to an extrinsic deformity which suggests the presence of an extrinsic mass (RPO & AP projections).

### Diagnosis

Common bile duct stenosis, secondary to extrinsic mass.

### Comment

Extrinsic mass is suggested here because the configuration of the narrow segment of proximal common bile duct is smooth and tapered and involves predominantly one aspect of the common bile duct, producing an eccentric indentation. Thus, intrinsic abnormalities of the common bile duct, such as stricture, calculus, and tumor, are all unlikely.

In this case, a large solitary calculus (2 cm in diameter) of the gallbladder was found at surgery. It had eroded through the gallbladder wall as well as the adjacent common bile duct wall. An inflammatory mass surrounding the stone accounts for the narrowing seen on ERC. The pathologic diagnosis was cholecysto-choledocho fistula, as a variant of Mirizzi syndrome where the calculus erodes the gallbladder wall and impinges on the common bile duct, but may or may not extend into the common bile duct itself[1]. In such cases as this the differential diagnosis should include carcinoma of the bile ducts or gallbladder, both of which produce masses extrinsic to the common bile duct.

### Reference

1. Clement AR, Lowman RM: The roentgen features of Mirizzi syndrome. *Am J Roent* 94: 480-83, 1965.

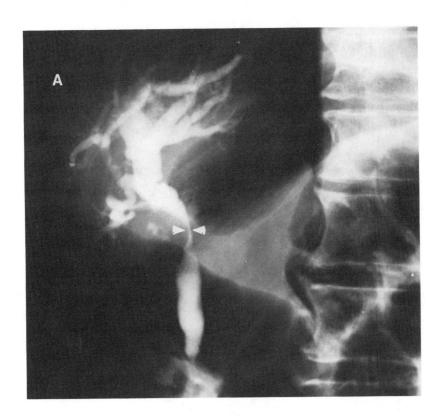

Case 14, Radiograph A

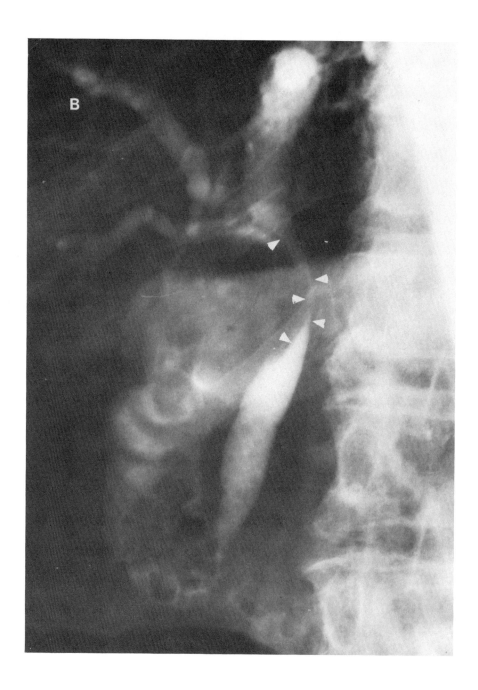

Case 14, Radiograph B

# IV

# Chronic Pancreatitis

# INTRODUCTION

In this section the full range of ductal expression of chronic pancreatitis is presented, including stenosis, dilatation, pseudocyst, and abscess, all associated with the disease in other than the acute phase. The pancreatic duct reflects the changes occurring within the parenchyma of the pancreas itself and, therefore, is not affected by pancreatitis in its early phase of development. As the disease process extends to include the duct and side branches they gradually show wall irregularity and contour change initially, then irregular dilatation, and finally, beading and lack of the normal delicate progressive tapering from the ampulla proximally. The identification of any one of these abnormalities permits diagnosis. Further, since segmental involvement of the pancreatic duct is characteristic of pancreatitis, any of these changes seen, either throughout or in a single segment, also permits diagnosis of chronic pancreatitis. ERCP is, therefore, of considerable value in diagnosing moderate and severe stages of chronic pancreatitis, with limited capability of identifying the earliest phase.

In the extensive pathological process which chronic pancreatitis encompasses, pseudocyst and/or abscess are representative, generally, of advanced disease and indicate gross forms of tissue destruction. ERCP is usually not required to establish the diagnosis of either of these, but rather to characterize the status of the pancreatic duct, since their presence generally will already have been demonstrated by ultrasound, CT, and more rarely, plain radiographs. Thus, those cases shown here as "unusual" are primarily the pseudocyst cases, where the intent was to determine locus and number of stenoses along the course of the duct, the presence of calculi, and the extension (as nearly as safely possible) of the pseudocyst. ERCP functions here to identify the remediable condition insofar as surgical management is concerned.

The single visual characteristic which identifies the presence of chronic pancreatitis is distortion of the normal duct architecture such that the overall delicate tapering and arborization are lost, so that one sees instead irregular margins, stubby side branches, and alternating stenosis and dilatation. Given an affirming clinical setting, the presence of any combination of these features is diagnostic of chronic pancreatitis, even when they affect only a single region (head, body, or tail). (See Case 17.)

## CASE 15
### (Accompanying Radiograph A)

### Clinical History

A 38-year-old white male presented with a 28-year history of alcoholism. He had multiple previous admissions for chronic pancreatitis. Eight years prior to this admission he had a vagotomy and pyloroplasty for gastric ulcer, and 3 years before had an exploratory laparotomy for GI bleeding. On physical examination the liver was enlarged and urine amylase was moderately elevated.

### Radiographic Findings

A. ERCP shows the pancreatic duct to be moderately dilated throughout its length with slight wall irregularity and caliber variation. There is an abrupt decrease in caliber in the tail segment of the duct. A few side branches are filled and these are stubby, straightened, and show wall irregularity and occasional beading.

### Diagnosis

Chronic pancreatitis, moderate, diffuse.

### Comment

This case represents an example of moderate chronic pancreatitis because of the amount of dilatation present throughout the duct without the presence of complications such as stricture, pseudocyst, or abscess. Were changes of similar severity seen to affect only a localized portion of the duct, nonetheless, the stage would be qualified as moderate. However, such a circumstance permits designation of that anatomic region as the specific site of involvement. Thus, dilatation of only the tail portion of the pancreatic duct of the same degree seen in this case would be diagnosed: chronic pancreatitis, moderate, involving the tail.

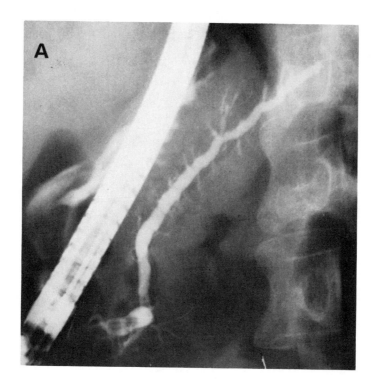

Case 15, Radiograph A

## CASE 16
### (Accompanying Radiographs A & B)

### Clinical History

A 31-year-old white female presented for her thirtieth admission on this occasion with abdominal pain, diarrhea, nausea, and vomiting. She had had a history of psychiatric problems and was a known laxative abuser. On the current admission there was evidence of phenolphthalein in the stool, and she had symptoms of pancreatitis with a moderately elevated serum amylase. Serum potassium was low (2.2 mg/100 ml). Upper GI series with small bowel, barium enema, and oral cholecystography were all normal. ERCP was performed to clarify the differential diagnosis which included: laxative abuse; pancreatic cholera; Zollinger-Ellison syndrome; and chronic pancreatitis.

### Radiographic Findings

A. ERCP shows segmental dilatation and irregularity of the proximal and middle
& thirds of the pancreatic duct, with relative narrowing at the junction of the
B. middle and proximal thirds (arrows).

### Diagnosis

Chronic pancreatitis, moderate, involving the tail and body.

### Comment

The tail and body may be specified as the exclusive sites of involvement by chronic pancreatitis because of the presence of pancreatic duct dilatation at these locations (see Case 15). What is described as the "relative narrowing" at the junction of the middle and proximal thirds is, nonetheless, normal duct which appears narrow because it occurs between two dilated segments. Were it truly narrow, the diagnosis would include *stenosis* (or, rather, in this case, *stricture*). Diagnosis of stenosis or stricture requires an assessment as to what the duct caliber would normally be in a given patient in the affected region, and it is based on reference to visualized normal duct. In this case, if one assumes the caliber of the pancreatic duct in the head is normal, by extrapolation the narrowed segment would be of appropriately reduced caliber for that location and, thus, normal.

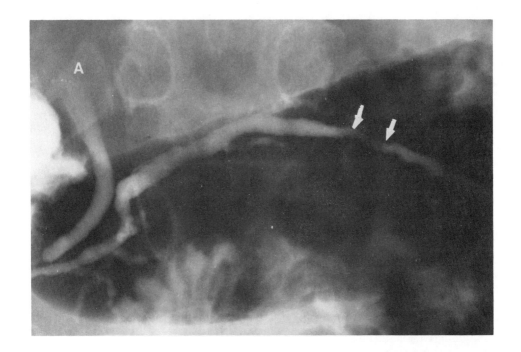

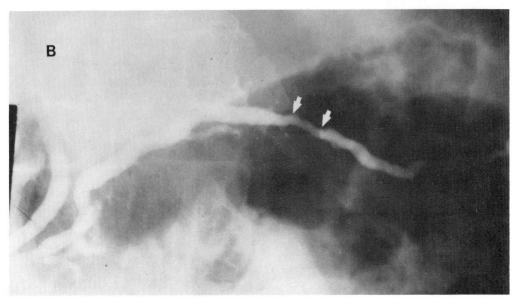

Case 16, Radiographs A & B

**CASE 17**
(Accompanying Radiographs A & B)

**Clinical History**

A 23-year-old black female presented with a lengthy history of abdominal pain. Multiple upper GI series prior to the current admission were all negative. Physical examination on admission was negative as well, and the patient denied alcohol abuse.

**Radiographic Findings**

A. ERCP shows complete filling of the pancreatic duct as well as of several small
& side branches. In its middle third (arrows) it is moderately narrowed relative to
B. the caliber throughout, and the side branches are straightened.

**Diagnosis**

Chronic pancreatitis, moderate, middle third.

**Comment**

Features which permit diagnosis here are: a) the pancreatic duct is completely filled as judged by visualization of the tail portion and of several side branches throughout; b) the duct is narrowed disproportionately, in its middle third; and c) side branches which are filled in the middle third are straightened.

Diagnosis demands complete filling of the pancreatic duct which is not infrequently accompanied by some side-branch filling as well. This case represents a good example of that degree of side-branch filling which is considered appropriate with good technique. As discussed in Case 16, narrowing, or rather tapering, is a relative feature varying from normal to abnormal in degree and it is determined by comparison with a known normal segment. In this case, the caliber of the distal portion is within normal range, and the normal caliber reduction to the middle third exceeds the standard.

Normal side branches seen in the head (Radiograph A) show progressive tapering, are 1 to 3 mm in length, 1 mm in caliber, and are slightly sinuous in their course. In the middle third of the pancreatic duct, however, side branches are straightened and lack the usual random sinuosity. This occurs because edema and fibrosis in the surrounding parenchyma distort normal acinar and lobular architecture. Stubby or beaded side branches (Cases 15 and 18) are seen also for the same reason and appear to represent a more advanced stage of pancreatitis. With regard to duct tapering in general, and in the prone position particularly, the spine may contribute, through extrinsic compression, some slight narrowing in the middle third of the pancreatic duct. For this reason there are caliber reductions slightly in excess of the standard 1 mm from head to body which may occur, and which should be regarded as normal.

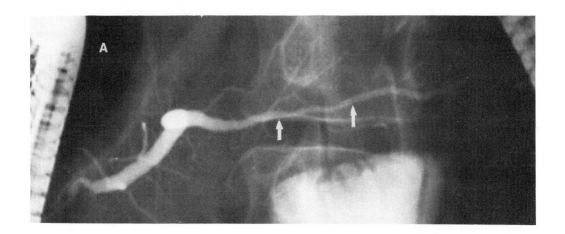

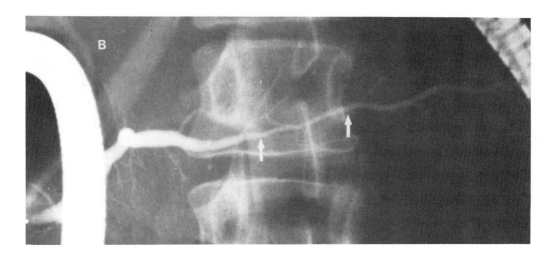

Case 17, Radiographs A & B

## CASE 18
### (Accompanying Radiographs A & B)

### Clinical History

A 34-year-old black male presented with a 4-year history of pancreatitis associated with multiple previous hospital admissions. He had had persistent epigastric pain for 1 month prior to the current admission. Serum amylase was moderately elevated. Upper GI series, oral cholecystography, and ultrasonography were all normal.

### Radiographic Findings

A. The pancreatic duct shows a stenosis within the head (arrowheads) with proximal dilatation and a second stenosis at the junction of the proximal and middle thirds (arrowheads). Several stubby, slightly dilated side branches are seen in the middle third.

B. Adjacent to the stenosis in the distal pancreatic duct there is also stenosis of the distal third of the common bile duct (arrowheads) with slight proximal dilatation.

### Diagnosis

Chronic pancreatitis, diffuse, with multiple stenoses of the pancreatic duct and common bile duct.

### Comment

The presence of segmental dilatation as well as side-branch stubbiness establish the diagnosis of chronic pancreatitis. Further, the additional characteristic feature present is stenosis: two areas in the pancreatic duct, and one in the distal common bile duct. Significantly as well, all of these persisted throughout the examination. Therefore, the diagnosis of stenosis is warranted because of the nontransient quality of these narrowed areas. They might also have represented strictures, that is, narrowing due to fibrosis resulting from chronic pancreatitis. However, because the etiology cannot be ascertained at ERCP, the formal diagnosis must be limited to: *stenosis, presumed stricture.* In fact, in this case no scarring of the common bile duct was discovered at subsequent surgery, and the stenosis was due to edema, and is, thus, the appropriate descriptive term.

With regard to the clinical application of the finding of stenosis, the capability of ERCP to identify multiple and/or severe stenoses of the duct represents a direct aid to therapy in the selection and planning of appropriate treatment. In this case, for example, partial pancreatic resection was subsequently performed with Puestow pancreatojejunostomy and Roux-en-Y procedure, and common bile duct stenosis was confirmed as due to edema of the head of the pancreas.

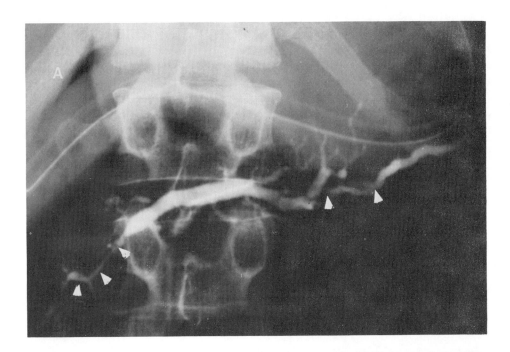

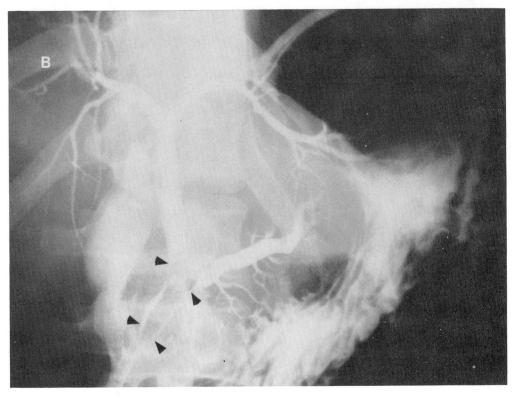

Case 18, Radiographs A & B

The differential in a case with terminal involvement of adjacent segments of both pancreatic and common bile ducts (the so-called *double duct sign*) should include carcinoma as well as chronic pancreatitis. Those features specifically characteristic of tumor will be discussed in Section V. Here, although the distinctive features of chronic pancreatitis are present and permit such a diagnosis, the possibility of carcinoma as a second pathology is not eliminated thereby. However, the radiologic impression here considers the adjacent involvement of the pancreatic and common bile ducts in their terminal portions to be etiologically integral to processes related to chronic pancreatitis[1,2]. Thus, although carcinoma is dominantly associated with the double-duct sign, this is not an exclusive relationship[3], as should be clear on the basis of its pathogenesis (see Case 33, Comment).

## References

1.  Wisloff F, Jakobsen J, Osnes M: Stenosis of the common bile duct in chronic pancreatitis. *Br J Surg* 69:52-54, 1982.

2.  Gregg JA, Carr-Locke DL, Gallagher MM: Importance of common bile duct stricture associated with chronic pancreatitis. Diagnosis by endoscopic retrograde cholangiopancreatography. *Am J Surg* 141:199-203, 1981.

3.  Plumley TF, Rohrmann CA, Freeny PC, Silverstein FE, Ball TJ: Double duct sign: reassessed significance in ERCP. *Amer J Roentgenol* 138:31-35, 1982.

# NOTES

## CASE 19
### (Accompanying Radiograph A)

### Clinical History

A 28-year-old black male presented with a history of chronic alcoholism and episodes of abdominal pain. Physical examination and LFT were normal.

### Radiographic Findings

A. ERCP shows moderate dilatation throughout the pancreatic duct, especially in the middle third, with substantial stenosis of the tail portion. The few side branches that do fill are straightened and stubby.

### Diagnosis

Chronic pancreatitis, moderate, diffuse.

### Comment

Dilatation of the pancreatic duct caused by chronic pancreatitis, either segmental or generalized, occurs as a consequence of and to the degree that: a) the surrounding parenchyma in the head or body or tail may be atrophied; and/or b) there is stenosis or stricture resulting from scarring, usually extrinsic to the duct itself. Because stenosis may not only exacerbate existing atrophy but may also produce it, it is essential in terms of clinical management to establish its *absence* as well as its presence. Careful search must be made from head to tail, with particular scrutiny of the distal portion of the pancreatic duct since proximal stenosis has less geographic significance.

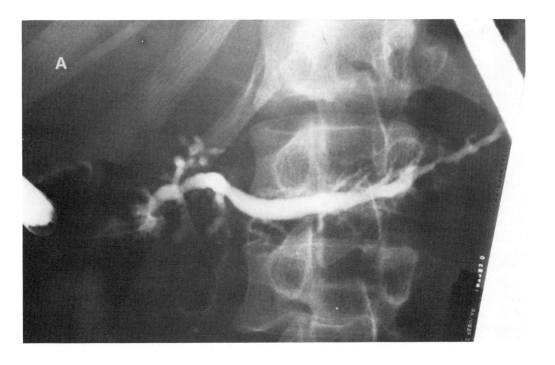

Case 19, Radiograph A

## CASE 20
### (Accompanying Radiographs A & B)

### Clinical History

A 50-year-old black male presented with burning epigastric pain. He had had a lengthy history of alcoholism with associated chronic pancreatitis for which this was the tenth admission. He had been treated previously for a pseudocyst, and prior surgery included cholecystectomy and sphincteroplasty 4 years before the current admission. ERCP was performed to rule out recurrent pseudocyst.

### Radiographic Findings

A. There is massive dilatation of the main pancreatic duct and also of an accessory
& duct in the head. The dilatation involves all regions (including the tail which is
B. foreshortened in this RPO projection). A few stubby branches are visualized throughout.

### Diagnosis

Chronic pancreatitis, severe, diffuse.

### Comment

The degree of dilatation present here directly implies advanced disease with extensive parenchymal destruction. Yet, as with Case 19, no distal stenosis is present. Thus, not only are the changes seen here entirely a reflection of glandular atrophy, but demonstrate as well that fibrosis associated with stenosis is a feature which accompanies dilatation only when it occurs directly adjacent to the pancreatic duct regardless of the amount of parenchymal destruction.

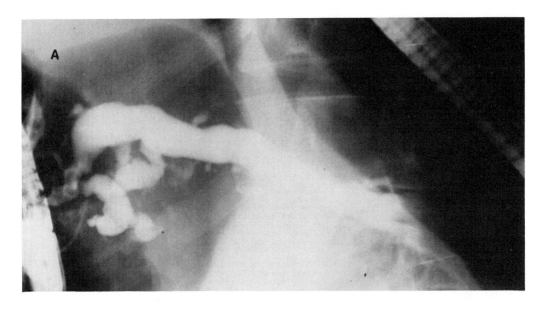

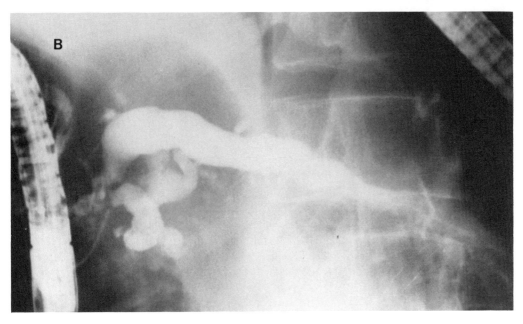

Case 20, Radiographs A & B

## CASE 21
### (Accompanying Radiographs A-C)

### Clinical History

A 40-year-old female presented with a 10-year history of alcoholism with associated chronic episodes of abdominal pain of unknown etiology. On current admission the physical examination was negative and serum amylase was only slightly elevated.

### Radiographic Findings

A. The pancreatic duct is normal to the junction of the proximal and middle thirds at which point there is severe stenosis (arrow).

B. There is moderate ectasia and tortuosity of the proximal portion of the duct which contains multiple, small, radiolucent calculi (arrowheads).

C. This magnified view of the tail emphasizes the localized nature of the stenosis (arrows), and also shows the calculi and debris in the dilated proximal portion of the pancreatic duct.

### Diagnosis

Chronic pancreatitis, moderate, localized to tail, with associated stenosis and calculi.

### Comment

In contrast to Case 20, this pancreatic duct, which is entirely normal in body and head, is stenosed at the tail with secondary calculus formation; the stage of disease is less advanced overall than in Case 20. The remarkable feature here is the localized character of the pancreatitis, limited as it is to one segment. It is interesting to speculate whether the stricture, due to pancreatitis or such a process as calculus disease, produced proximal duct dilatation; or, whether recurrent pancreatitis led to atrophy, duct dilatation, and associated stricture (since calculus formation commonly accompanies stricture).

Nagata et al.[1] have studied the use of ERCP as a method of determining the natural history of pancreatitis and claim that there is a distinction between alcoholic and nonalcoholic subjects in: a) the extent of ERCP changes; and b) the amount of progression seen with time. There are certain limitations to this approach in that the ability of ERCP to quantify the early changes of pancreatitis has not been defined.

### Reference

1. Nagata A, Homma T, Tamai K, Ueno K, Shimakura K, Oguchi H, Furuta S, Oda M: A study of chronic pancreatitis by serial endoscopic pancreatography. *Gastroenterology* 81:884-891, 1981.

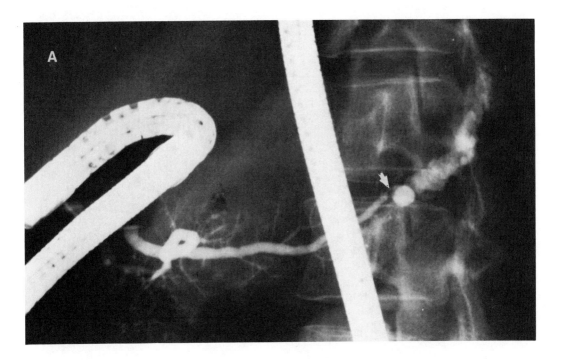

Case 21, Radiograph A

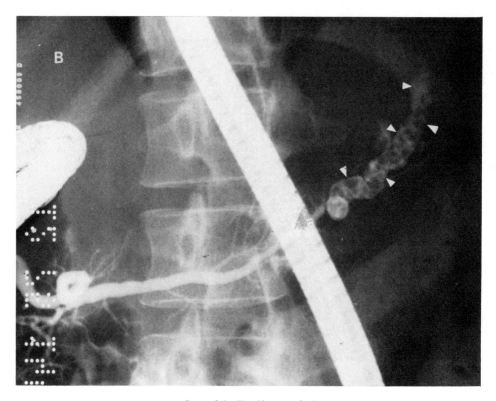

**Case 21, Radiograph B**

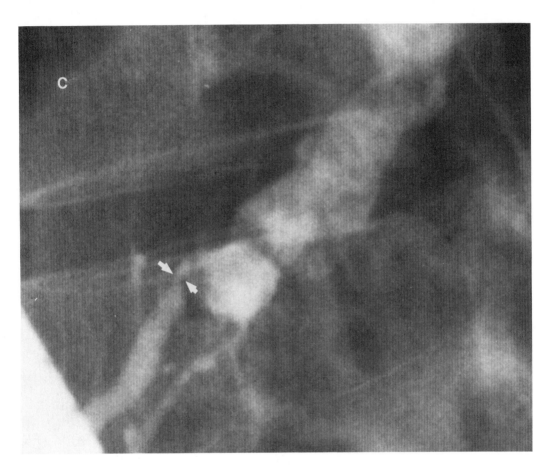

Case 21, Radiograph C

## CASE 22
### (Accompanying Radiographs A & B)

### Clinical History

A 56-year-old black male presented with a history of alcoholism and gastric ulcer for many years as well as a 5-year history of intermittent abdominal pain. For 1 month prior to the current admission the pain had radiated to the back and he had sustained a 25-pound weight loss during that period. On admission examination he had only slight upper abdominal tenderness. After 2 weeks in the hospital, he developed a high, spiking fever with a WBC of 17,000 and urine amylase twice the normal level.

### Radiographic Findings

A. ERCP shows normal pancreatic duct in the head and distal body (arrowheads) with considerable duct-caliber variation.

B. There is filling of an irregular cavity about 3 cm in diameter in the mid-body (1), and saccular ectasia of the pancreatic duct proximally (2) on both Radiographs A and B. Note the endoscopy catheter within the common bile duct (3).

### Diagnosis

Chronic pancreatitis, diffuse, severe, with abscess in the body of the pancreas (confirmed at surgery).

### Comment

Radiologically, neither pseudocyst nor abscess present distinctively. Both visualize as irregular collections of contrast material provided there is a communication with the pancreatic duct. The size for both varies in the extreme, from a few milliliters to several liters. The extent to which they are visualized is a factor dependent upon the volume of contrast injected and amount of debris they contain. There is no characteristic configuration of the margin for either. Thus, ERCP per se cannot distinguish pancreatic abscess from pancreatic pseudocyst. On the basis of appearance alone the irregular cavity seen in this case might well have been either. What permits a definitive diagnosis of abscess is the clinical history. Specifically, the presence of both high spiking fever and a markedly elevated WBC indicate the existence of an infectious process. Since the ERCP clearly demonstrates chronic pancreatitis as well as a structure (Radiograph B, 1) compatible with an abscess, pseudocyst may be excluded.

This case highlights the relationship in film interpretation between the radiologist and the range of objective data available to him which enables diagnosis. Here, for example, the "Findings" intend to no more than list the significant features in as generic terms as possible with the exception of recognizable anatomic structures. Thus, "cavity," not "abscess." Thus, "saccular ectasia," not "chronic pancreatitis." Interpretation is a synthetic process, occurring between findings and diagnosis, where all data, clinical and radiologic, is assembled and analyzed. Here this process took into consideration all of the features on ERCP in addition to clinical features such as abdominal tenderness, spiking fever, and WBC elevation. The pivotal quality of the WBC here in narrowing the differential to abscess demonstrates that ERCP maintains a closer radiologic relationship with clinical settings than is usual.

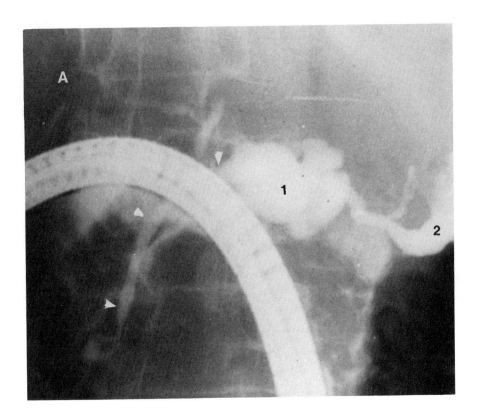

Case 22, Radiograph A

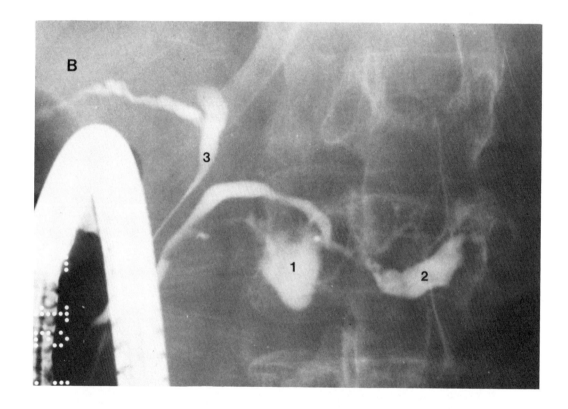

Case 22, Radiograph B

# NOTES

**CASE 23**

(Accompanying Radiographs A & B)

### Clinical History

A 38-year-old white male presented with a 2-year history of hepatitis and associated renal and cardiac failure. There were several prior admissions for abdominal pain. Serum amylase on the current admission was initially normal, and later increased moderately. Oral cholecystography, intravenous cholangiography, celiac arteriography, and upper GI series were all normal.

### Radiographic Findings

A.  The distal 4 cm of the pancreatic duct are normal (arrowheads). There is an area of stenosis in the head (arrow) with progressive accumulation of the opaque material in amorphous, round, loculated areas which replace much of the body and the tail.

B.  These amorphous collections are especially well seen on the delayed films at 30 minutes (arrowheads).

### Diagnosis

Chronic pancreatitis with stenosis of the distal pancreatic duct and probable massive dilatation at the body and tail.

### Comment

The differential in this case included abscess or pseudocyst, and pancreatic duct ectasia. Note that the large irregular structures opacified on Radiograph B are, in their irregular, undefined margins, similar to the cavity structure seen in Case 22, Radiograph B, 1. In that case duct dilatation was eliminated from the differential consideration because complete, intact pancreatic duct could be seen. Therefore, the cavity had to have been a separate structure and this implies the presence of pseudocyst or abscess. In this case, however, such a discrimination cannot be made because there are no normal landmarks beyond the point of stenosis. Thus, it is impossible to determine whether the opacified structure is in fact duct, pseudocyst, or abscess; or whether normal or abnormal duct is present but obstructed and hence not visualized. The only objective ERCP findings which may be used here consist of the general contour and orientation of the opacified structures which follow, albeit amorphously, that of pancreatic duct within a pancreas. The diagnosis in this case must be based, therefore, on probability and differential considerations can be ranked accordingly. Thus, chronic diffuse pancreatitis with massive duct ectasia seem more likely than the other differential possibilities.

This case demonstrates the value of even a single abnormality (cavity) or land-mark (contour and orientation) as a point of departure in making a diagnosis. Further, the therapeutic importance of identifying any abnormality is of itself substantial. Here, for example, any of the differential considerations were sufficient to necessitate surgery. In fact, in this case the autopsy performed 2 weeks after surgery confirmed, point for point, the diagnosis on ERCP.

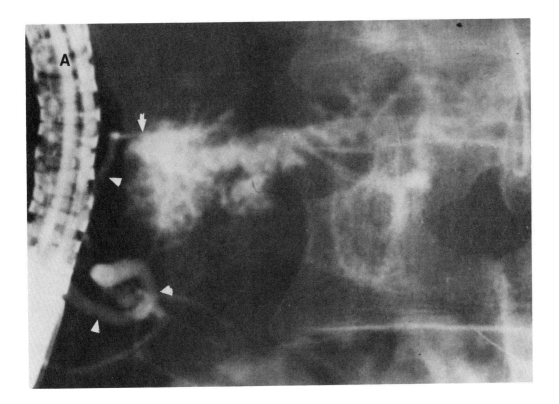

**Case 23, Radiograph A**

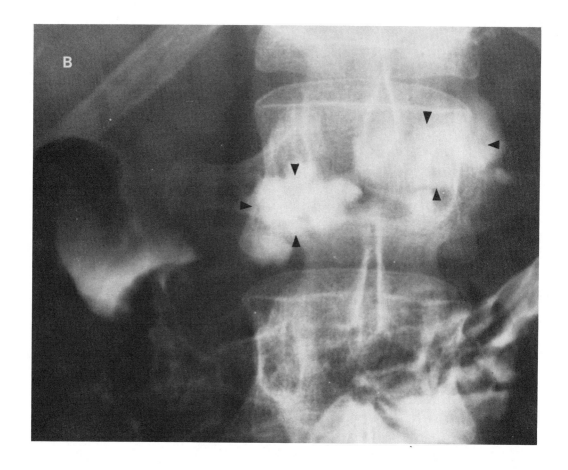

**Case 23, Radiograph B**

# NOTES

## CASE 24
### (Accompanying Radiographs A-D)

### Clinical History

An 18-year-old white male had an episode of abdominal pain, emesis, and diarrhea over a 2-day period prior to the current admission, and following excessive alcohol intake. On admission he was febrile, serum amylase was moderately elevated, and there was no jaundice.

### Radiographic Findings

A. There is severe stenosis of the pancreatic duct in the head about 4 cm proximal to the ampulla (arrow). The opaque then fills and defines an amorphous, round cavity in the body of the pancreas (arrowheads).

B. ERCP, in the LPO view, better defines the site of stenosis (arrow) which occurs between the normal pancreatic duct and the cavity.

C. With increasing filling there is progressive delineation of the cavity which is seen to occupy most of the body and tail (white arrowheads), and contains irregular filling defects. The pancreatic duct in the head is normal (black arrowheads). Contrast material in the stomach partially overlies the region.

### Diagnosis

Pseudocyst or abscess of the body and tail of the pancreas.

D. Subsequent intraoperative pancreatic cystogram shows a large bi-lobed structure involving the body and tail.

### Comment

As with Case 23, the differential includes duct ectasia, pseudocyst, or abscess. Duct ectasia can be eliminated readily because of the rounded configuration of the outline of the cavity, particularly well visualized by the stage of filling achieved at Radiograph C. Duct ectasia, with rare exception (Case 23), assumes a tubular, sausage-like shape (Cases 18 and 19), even with massive ectasia (Case 20). In discriminating pseudocyst from abscess, the irregular filling defects seen are not useful data since these may represent, equally, either debris (abscess) or multilocularity (pseudocyst). Nor is direct injection into the cavity during intraoperative cystography with demonstration of the cavity's bi-lobed quality particularly definitive. Here the differential was narrowed to a single definitive diagnosis only at surgery, where a large bi-lobed pseudocyst was found and cystogastrostomy performed.

These last three cases, Cases 22, 23, and 24, present examples of abscess, duct ectasia, and pseudocyst respectively. The significance of these, related to differential diagnosis, is not connected with the difficulty of discrimination among them, but rather that other pathology can be definitively *excluded.* Further, with regard to the clinical use of ERCP as a diagnostic study, its ability to narrow the differential to phenomena related to chronic pancreatitis is sufficient, since all are benign processes and all require surgical intervention.

Intraoperative pancreatic cystography is not a major diagnostic aid since, generally, the size of the cyst or abscess is unimportant in determining the surgical procedure per se. However, the determination of size, as well as the detection of tracking and sinus formation, may be useful information, nonetheless, in modifying the surgery. Therefore, because this has proven to be a safe procedure with careful technique, its use may be encouraged in cases of suspected pseudocyst or abscess.

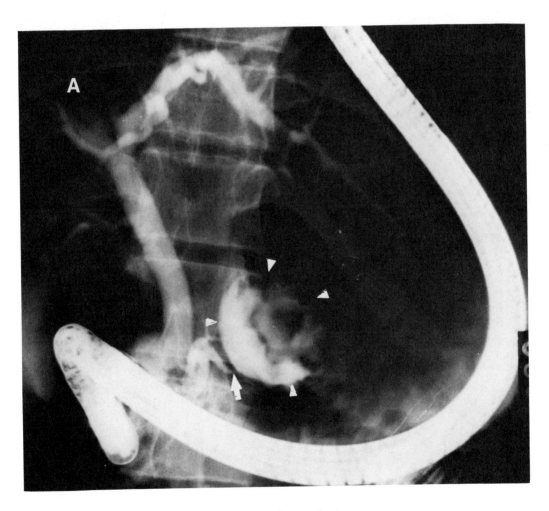

Case 24, Radiograph A

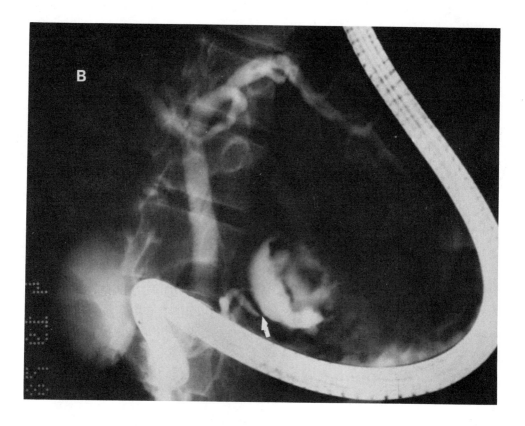

**Case 24, Radiograph B**

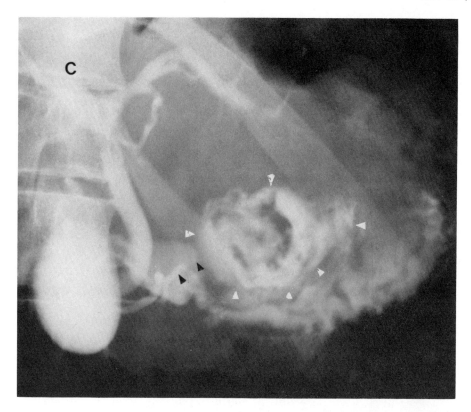

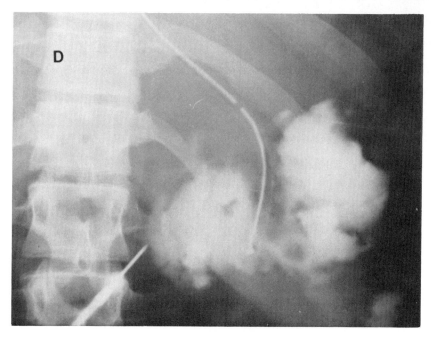

Case 24, Radiographs C & D

## CASE 25: UNUSUAL CASE
(Accompanying Radiographs A & B)

### Clinical History

Six months prior to the current admission, a 60-year-old white female presented with epigastric pain, nausea, and vomiting. Oral cholecystogram at that time did not visualize the gallbladder, and serum amylase was moderately elevated. With treatment her condition improved and she was discharged. She presented for this admission with recurrent abdominal pain and a palpable, tender, 5 cm epigastric mass. Serum and urine amylase were moderately increased.

### Radiographic Findings

A. ERCP shows stenosis of the pancreatic duct at the junction of the middle and distal thirds (arrow) where there is complete occlusion, best seen in this right lateral view which projects the pancreatic duct over the common bile duct.

### Diagnosis

Chronic pancreatitis with occlusion of the pancreatic duct at the junction of body and head.

B. Intraoperative cystogram, at laparotomy, shows a pseudocyst (arrows) involving the body of the pancreas. This was injected with contrast material which defines its size. It communicates with the duodenum via the pancreatic duct, resulting in partial opacification of the bulb and C loop.

### Comment

Findings of localized obstruction at the junction of the middle and distal thirds of the pancreatic duct were in themselves sufficient for appropriate therapy, that is, laparotomy and exploration of the pancreas. The operative cystogram, although not integral to surgical management, nonetheless was helpful in elucidating the pathophysiology here since it established that the process involved was a pseudocyst communicating with the pancreatic duct. In that regard, this case is an excellent example of a benign process causing apparent occlusion of the pancreatic duct. Most likely, the occlusion seen at ERCP resulted from the inflammatory process adjacent to the pseudocyst. Finally, one may also consider that this produced a valve-like obstruction which permitted flow in one direction only.

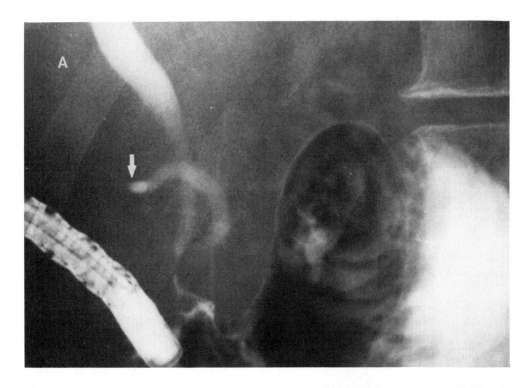

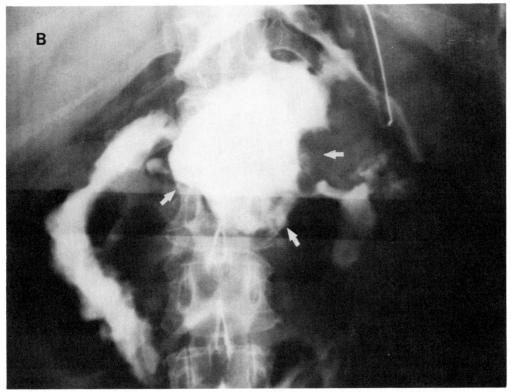

Case 25, Radiographs A & B

## CASE 26: UNUSUAL CASE
### (Accompanying Radiographs A-C)

### Clinical History

A 47-year-old white male presented with a history of alcoholism for many years associated with chronic pancreatitis. For about 2 weeks prior to the current admission he had had nausea, vomiting, and abdominal pain relieved by flexion of the hips. He had suffered a 40-pound weight loss during a period of 2 months prior to this admission. On physical examination he was cachectic without jaundice, and there was diffuse tenderness of the upper abdomen with a palpable mass in the left upper quadrant.

### Radiographic Findings

A. Upper GI series, left lateral view, shows a collection of gas bubbles in the soft tissues posterior to the body of the stomach (arrows) which involves an extensive area in the retroperitoneal space.

B. Upper GI series, an erect projection of the stomach, shows again the multiple small gas bubbles with a single large collection demonstrating an air-fluid level (arrows). The lesser curve of the stomach is indented and displaced to the left.

C. Intraoperative pancreatogram, done by direct catheterization of the pancreatic duct at surgery, demonstrates an irregularly narrowed pancreatic duct throughout the body (arrows) as well as an irregular bi-lobed collection of contrast material which fills the region of the tail (arrowheads).

### Diagnosis

Chronic pancreatitis, diffuse, with pseudocyst of tail (confirmed at surgery).

### Comment

This case is a unique example in this series of a bi-lobed pseudocyst at its most grotesque severity. Note that the gas bubbles in the soft tissues of the retroperitoneal space have coalesced to form a single large cavity within which there is an air-fluid level. When there is uncertainty with regard to the presence of small gas bubbles (since diagnosis of their presence must imply the existence as well of an infectious process) it is especially informative to obtain erect or decubitus views which utilize a horizontal x-ray beam and, thus, will demonstrate any confirming air-fluid levels if these are present. The differential in this case might consider perinephric abscess and other infectious processes; however, these are much rarer and are excluded here by the nature of the clinical history.

The intraoperative pancreatogram in this case demonstrated the communication between the pancreatic duct and the pseudocyst. While this information may not be diagnostically essential in such a case as this one, it is useful to the surgeon in planning operative procedures. For example, in this case the size demonstrated on pancreatogram led to a decision to perform subtotal pancreatectomy.

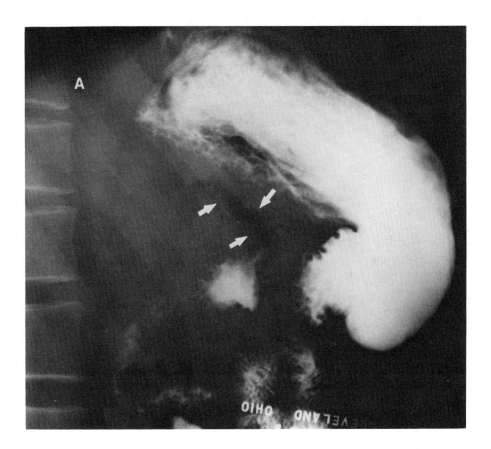

Case 26, Radiograph A

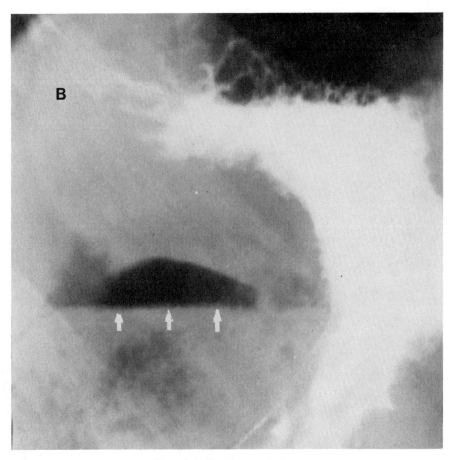

Case 26, Radiograph B

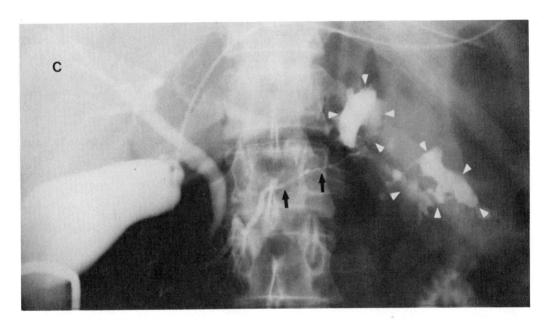

Case 26, Radiograph C

## CASE 27: UNUSUAL CASE
(Accompanying Radiographs A-D)

### Clinical History

A 36-year-old black male with a history of alcoholism for many years presented with abdominal pain, weight loss, jaundice, and right upper quadrant tenderness. The serum amylase was moderately increased.

### Radiographic Findings

A. Upper GI series shows deformity of the inner aspect of the second portion of the duodenum suggestive of an extrinsic process (arrows).

B. ERCP shows stenosis of the distal 5 cm of the pancreatic duct with a convex
C. upward sweep suggesting a mass in the head of the pancreas inferior to the pan-
& creatic duct. Proximal to the stenosis, the pancreatic duct (1) is markedly dila-
D. ted, shows lack of side-branch filling, and demonstrates irregular caliber variation. Note also the filling of an accessory duct (2), and a duodenal diverticulum (3).

### Diagnosis

Chronic pancreatitis, severe, diffuse with pseudocyst of the head (confirmed at surgery).

### Comment

The C-loop deformity seen on the upper GI series is due primarily to the diverticulum of the second portion of the duodenum and is unrelated to the pseudocyst. While the dilatation of the pancreatic duct may be regarded as due to distal obstruction, the irregular caliber variation and lack of side-branch filling are more suggestive of preexisting chronic pancreatitis and, in this case, with a superimposed component of dilatation secondary to distal obstruction due to extrinsic processes. Thus, this case represents a rare example of extrinsic compression of the pancreatic duct due to a benign process. In this regard, it is especially noteworthy that there is extensive deformity and displacement of the distal pancreatic duct. Range of potential causes of the 5 cm smooth extrinsic compression include pseudocyst, abscess, and benign and malignant tumor. In a patient of this age, particularly with a history of known pancreatitis, in addition to the ERCP changes likely representing chronic pancreatitis, tumor was eliminated from consideration as rare. Abscess was eliminated also because of the absence of inflammatory changes (edema or spasm). In general, the differential diagnosis in unusual cases, where the unique features may not be definitively characterized, proceeds on the basis of consideration of those processes which are most probable.

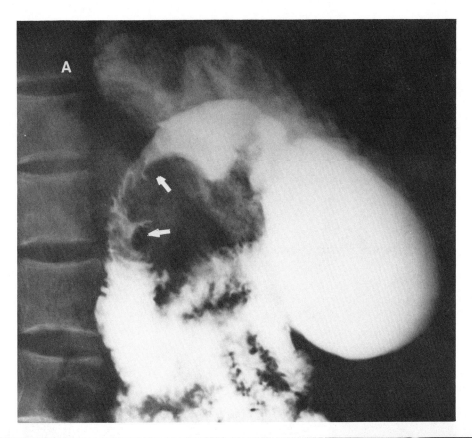

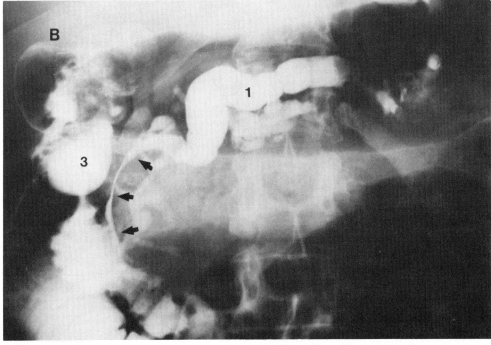

Case 27, Radiographs A & B

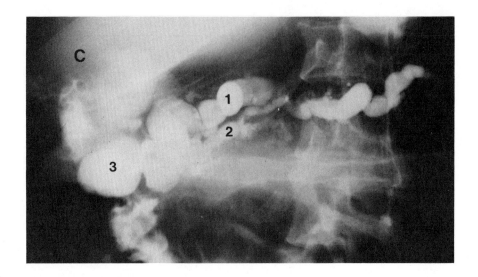

**Case 27, Radiograph C**

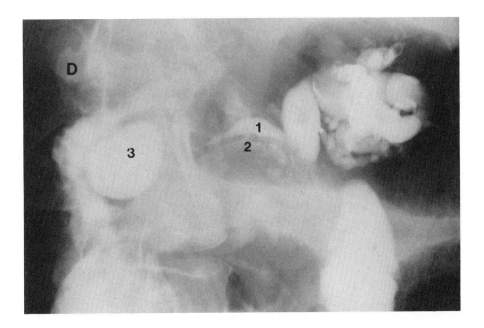

**Case 27, Radiograph D**

## CASE 28: UNUSUAL CASE
### (Accompanying Radiographs A-D)

**Clinical History**

A 28-year-old white male with a lengthy history of alcohol abuse presented with severe abdominal pain 24 hours after he had sustained direct trauma to the abdomen. (He stated that he had been kicked.) Physical examination revealed rare bowel sounds, diffuse tenderness, and involuntary spasm of the left upper and lower quadrants with voluntary guarding on the right but no ecchymoses. He underwent emergency exploratory laparotomy including lesser sac exploration. A large hematoma was noted over the surface of the pancreas which was quite easily palpable and considered to be normal and intact. Similarly, Kocher maneuver with bimanual palpation confirmed that the head of the pancreas was normal. Because the patient did not improve, and serum amylase (which had been moderately elevated on admission) continued to rise with associated persistent pain during the postoperative period, an upper GI series was performed 4 weeks after exploratory laparotomy.

**Radiographic Findings**

A.  Upper GI series shows displacement of the proximal jejunum and greater curve of the stomach, indicating a retroperitoneal process (arrowheads).

B.  ERCP, performed 1 week after upper GI series, shows a normal pancreatic duct up to the junction of the proximal and middle thirds (arrow), at which point there is extravasation of the contrast material (arrowheads).

C.  With continued injection, there is progressive accumulation of the contrast
&   material from the point of extravasation and extending toward the tail of the
D.  pancreas. This corresponds to the region of suspected mass demonstrated on Radiograph A (arrowheads).

**Diagnosis**

Pancreatic duct rupture secondary to direct trauma with pseudocyst formation (confirmed at surgery, where a large pseudocyst—about one liter—was found in the extreme left upper quadrant in the area of the eleventh rib). A Roux-en-Y enteroenterostomy was performed for drainage with eventual complete recovery.

**Comment**

The principal diagnostic feature here which permits diagnosis of pseudocyst is the extravasation of contrast material, which is distinctly cavity-like rather than spidery as produced by tumor, and which obviously originates directly from the duct. Further, it occupies an area which closely conforms with that seen as producing a mass effect in Radiograph A.

The particular usefulness of ERCP in diagnosing pancreatic trauma is clearly demonstrated by this case. Trauma to the pancreas was suspected on admission. The attending surgeon performed a meticulous laparotomy which included exploration of the lesser sac and anterior surface of the pancreas, as well as a Kocher maneuver. He nonetheless was unable to substantiate the presence of pancreatic injury. ERCP provided not only the evidence of duct injury, but its specific site and the complication of pseudocyst as well. It seems appropriate that ERCP be regarded as necessary to the complete evaluation of patients who may have sustained abdominal trauma and suspected pancreatic injury, especially when abdominal pain and amylase elevation persist[1].

Pancreatic injury as a result of blunt abdominal trauma is usually associated with automobile accidents, where the mechanism is often compression of the abdomen against the steering wheel.

Assault, falls, gunshot, and knife injuries are less commonly the cause of trauma.

### Reference

1.  Belohlavek D, Merkle P, Probst M: Identification of traumatic rupture of the pancreatic duct by endoscopic retrograde pancreatography. *Gastrointest Endosc* 24:255-256, 1978.

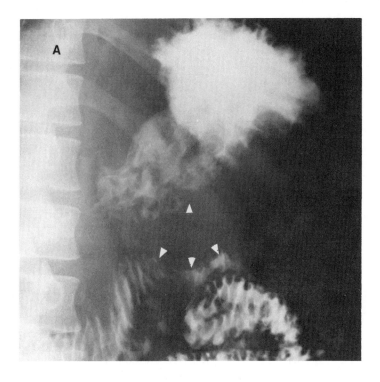

Case 28, Radiograph A

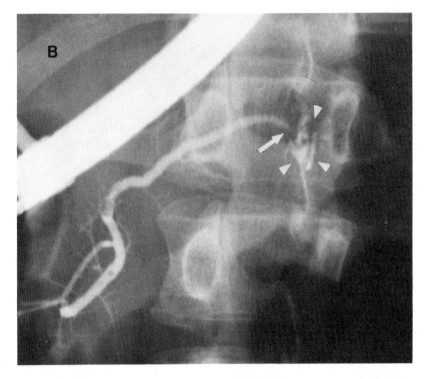

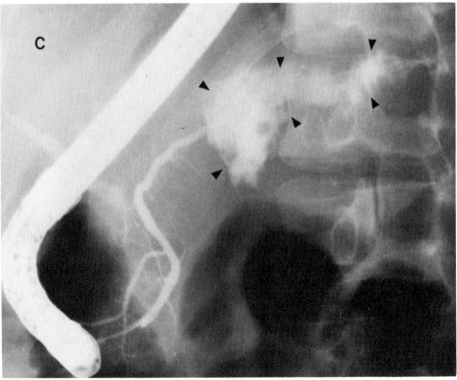

Case 28, Radiographs B & C

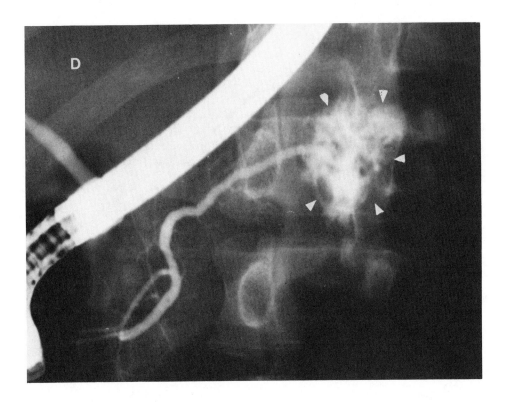

Case 28, Radiograph D

**CASE 29: UNUSUAL CASE**
(Accompanying Radiographs A-G)

## Clinical History

A 35-year-old white male presented with a history of chronic pancreatitis associated with alcoholism for 11 years, necessitating twelve hospital admissions. He had had epigastric and back pain for 2 months prior to the current admission.

## Radiographic Findings

A. The pancreatic duct is considerably dilated from ampulla to midbody at which
&  point there is progressive tapering to a site of complete obstruction (arrow).
B. Throughout, there is caliber variation and filling of several stubby side branches. Wire sutures, seen in Radiograph A, are from previous laparotomy. The following series of plain radiographs of the left upper quadrant were obtained prior to ERCP and show the development of progressive calcification in the tail of the pancreas over a period of approximately 10 months.

C. Normal baseline, 43 months prior to ERCP.

D. Small, single, early calcification, 10 months prior to ERCP.

E. Increase in size of previously noted calcification, 9 months prior to ERCP.

F. Multiple calcifications, more widely distributed, 7 months prior to ERCP.

G. Extensive calcification throughout the tail and proximal body of the pancreas, forming a cast of the pancreatic duct and side branches. This film was obtained at the time of ERCP. Note striking resemblance of the cast to a fish skeleton.

## Diagnosis

Chronic calcific pancreatitis.

## Comment

Subsequent to ERCP the patient underwent a 95% pancreatectomy, and histologic examination confirmed calcification within the pancreatic duct and parenchyma with marked atrophy of pancreatic tissue throughout. This case is a unique example of rapidly developing calcification within the main pancreatic duct, side branches, and parenchyma, with an equally unique distribution of calcification. ERCP demonstrates the finding of chronic pancreatitis in the head and distal body, and shows the pancreatic duct occlusion associated with chronic pancreatitis. One may speculate that the rate of calcification has been accelerated by virtue of the presence of pancreatic duct occlusion. Usually, calcifications associated with chronic pancreatitis are parenchymal, scattered, punctate to a few millimeters in size, and develop during a period of at least several years. When calcification occurs within a duct, it does so in calculus form, and rarely do these pack to form a cast.

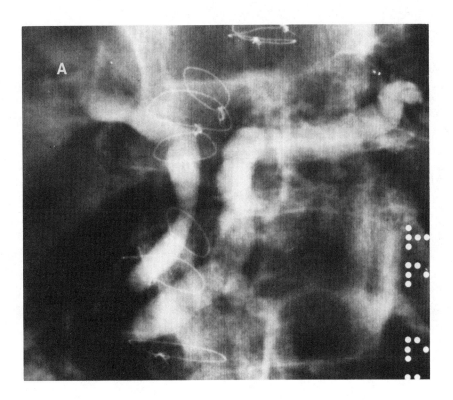

Case 29, Radiograph A

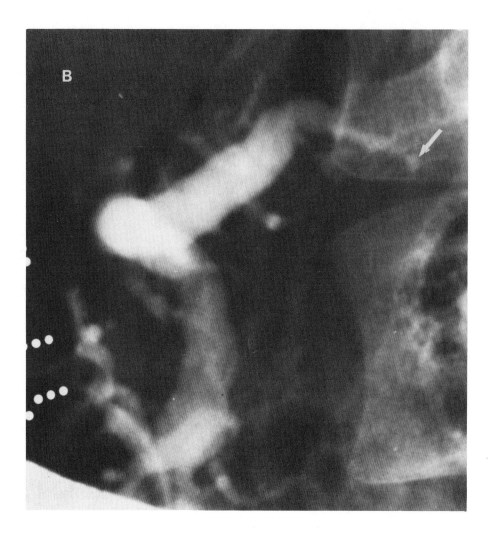

**Case 29, Radiograph B**

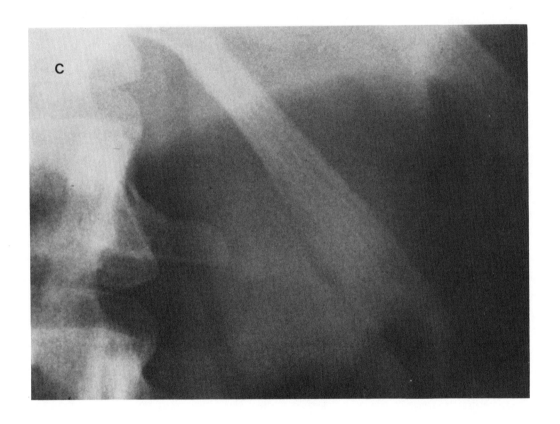

Case 29, Radiograph C

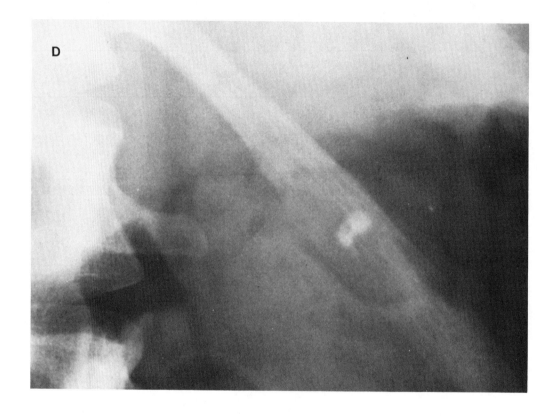

Case 29, Radiograph D

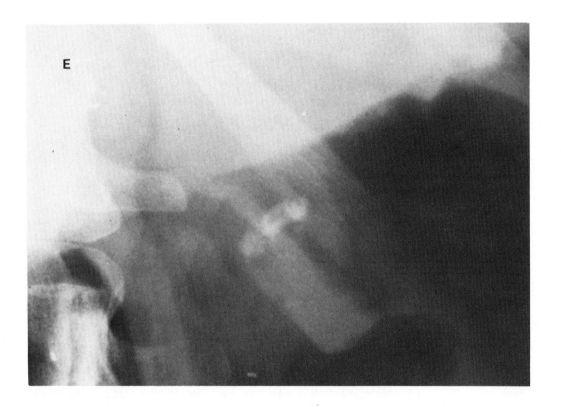

Case 29, Radiograph E

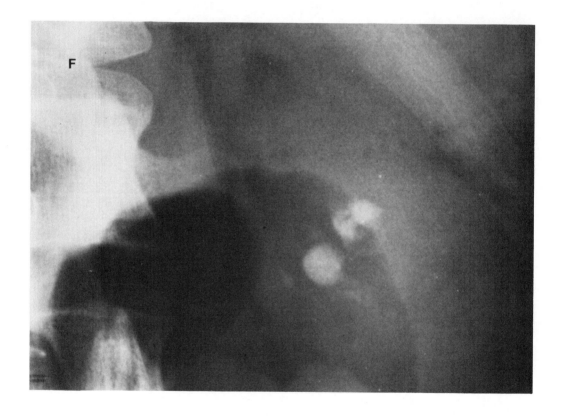

Case 29, Radiograph F

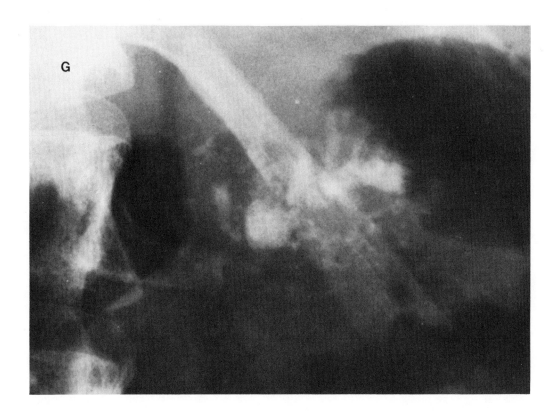

Case 29, Radiograph G

# V

# Carcinoma

# INTRODUCTION

As with chronic pancreatitis, carcinomas are diagnosed by ERCP in terms of the effect tumor masses in the gland parenchyma exert on the pancreatic duct. The chief expression of carcinoma is, as with chronic pancreatitis, stenosis, so that diagnosis depends on the ability to recognize and distinguish between frequently subtle characteristics of "malignant" stenosis.

There are, of course, a few obvious phenomena which may be associated with the presence of carcinoma: complete ductal obstruction, contrast material extravasation into a mass, and usually "double-duct" involvement. However, only the minority of these cases present with such patently characteristic signs of malignancy. For the majority, careful assessment of the nature of stenosis present is the basis of diagnosis.

So-called characteristic shapes of stenosis with, typically, beaked or rattail configurations are strongly suggestive of, but not definitive for, carcinoma. Simple, smooth tapered stenosis, on the other hand, is as likely to represent chronic pancreatitis as malignancy. Thus, the particular appearance of any stenosis must be interpreted with caution as to significance and in relation to associated signs and symptoms in the individual case. More than is called for by any other section of this book, the cases of carcinoma illustrate the need for a tentative, fine-tuned approach to analysis of detail presented on the radiograph.

**CASE 30**
(Accompanying Radiographs A & B. Autopsy Specimen C)

### Clinical History

A 63-year-old white male presented with a history of sudden 30-pound weight loss, but was otherwise asymptomatic.

### Radiographic Findings

A. Liver-spleen scan shows a solitary defect in the right lobe of the liver (arrow). A needle biopsy yielded adenocarcinoma.

B. ERCP, performed in search of a primary carcinoma, shows complete filling of the pancreatic duct in the body and the head with visualization of many side branches. There is an abrupt termination of pancreatic duct near the tail (arrow) rather than the smooth, tapering reduction in caliber which is normal in this region.

### Diagnosis

Carcinoma of the pancreas, tail.

### Autopsy

C. Autopsy specimen, 2 weeks after surgery, shows the pancreas and spleen viewed posteriorly. There is a 2 cm diameter carcinoma in the tail of the pancreas (white arrowheads) which occludes the pancreatic duct (arrow). Proximal to this point the duct, which is opened, is normal (black arrowheads). The pathologic diagnosis confirmed a primary carcinoma of the pancreas with metastasis to the liver.

### Comment

As has been previously discussed, nonfilling of any single segment of the pancreatic duct in the presence of careful technique must be regarded as diagnostically significant. The precise implication of nonfilling will be determined largely by the specific clinical circumstances. That is, the configuration of the occluded pancreatic duct (as in Radiograph B) does not permit, of itself, distinction between tumor, calculus, or stricture. However, the fact that this occurs in a geriatric patient here (where the incidence of carcinoma is greater) in the presence of sudden, significant weight loss, is indeed telling. If, for example, this were a 35-year-old patient who had sustained weight loss due to vomiting, the link with carcinoma would be far more tenuous. Needle biopsy of the suspected area might easily be performed for cytologic confirmation of a primary carcinoma when metastasis has not already been established as it was in this case.

Finally, in relation to carcinoma, it should be noted that it is not possible to differentiate by ERCP between primary and metastatic tumors involving the pancreas[1].

## Reference

1. Swensen T, Osnes M, Serck-Hanssen A: Endoscopic retrograde cholangiopancreatography in primary and secondary tumours of the pancreas. *Brit J Radiol* 53:760-764, 1980.

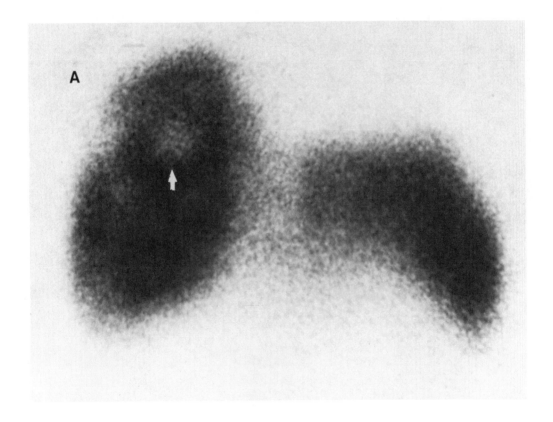

Case 30, Radiograph A

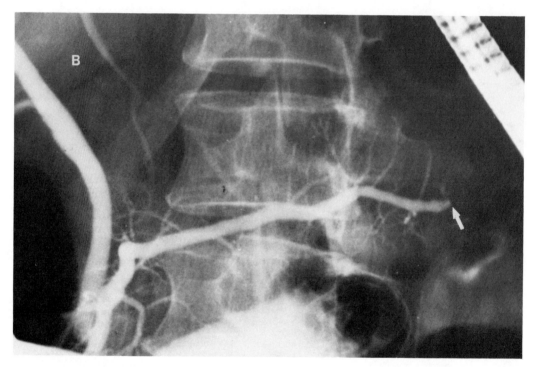

**Case 30, Radiograph B**

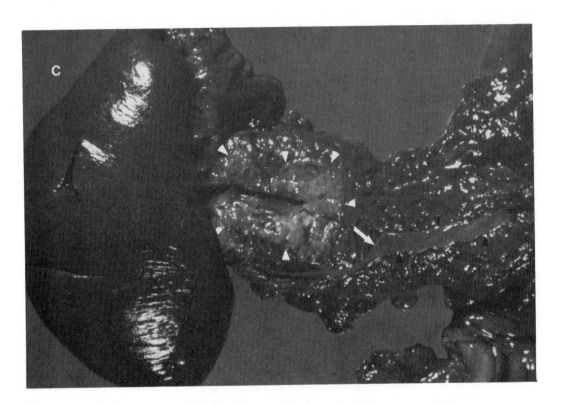

Case 30, Autopsy Specimen C

CASE 31
(Accompanying Radiograph A. Autopsy Specimen B)

## Clinical History

A 28-year-old black female presented with a 4-month history of generalized pruritus, easy bruising, and fatigability. About 1 month prior to the current admission she noted darkened urine, lightening of stools, and scleral icterus. On physical examination the bilirubin was markedly elevated. The common bile duct could not be cannulated on attempted ERCP, although the pancreatic duct was entered and appeared normal.

## Radiographic Findings

A. Percutaneous transhepatic cholangiogram, performed immediately following the attempted ERCP, shows marked dilatation of the common bile duct with a smooth, rounded termination. There is complete obstruction at this point which is several centimeters from what would be the normal location of the ampulla (arrowheads). The gallbladder (1) is visualized and no stones are seen within it or the intrahepatic branches of the biliary tree.

## Diagnosis

Common bile duct obstruction, complete, probably due to stricture.

## Autopsy

B. Autopsy specimen (posterior view) following the patient's demise after surgery shows a 2 cm, hard tumor mass in the head of the pancreas (black arrowheads) located on the posterolateral aspect of the common bile duct (in the region adjacent to the duodenal bulb and proximal second portion of the duodenum). This produced extrinsic compression of the common bile duct with complete occlusion. Note that the black ligature passes through the open distal common bile duct to lie free in the second portion of the duodenum (white arrow).

## Comment

In the case of obstruction, most significant to diagnosis is the demonstration of obstruction itself, as well as the level at which it is observed to occur. The particular configuration of the contrast material column in the distal common bile duct—smooth tapered, convex meniscus, concave meniscus, or rattail—is frequently concordant with the pathologic diagnosis. However, nearly as often one sees no correlation between the contour of the termination and the pathology. For example, accepted criteria in this case would imply the presence of a simple stricture (also see Case 32). While stricture was, indeed, the diagnosis made here, it was one based as much on the clinical data as on the visualization of obstruction per se. That is, given symptoms suggestive more of the presence of stricture or calculus (particularly in a patient as young as this one), carcinoma must be regarded as highly unlikely, rare—and in fact it is here. The tumor found, which was not the predictable adeno-carcinoma, could not be classified beyond a confirmation of malignancy, presumed to arise from pancreatic ducts.

This case is an excellent demonstration of the influence of the "operator error factor." That is, when the common bile duct cannot be cannulated in the context of good technique as well as consistent operator success in other difficult cases, lack of success (occasional) should at least be considered as due to altered anatomy. In this case, for example, the tumor mass so deformed the distal common bile duct and its relative anatomical orientation as to make cannulation mechanically impossible. Because operator skill was respected at the time of ERCP, complications which may have occurred as a result of forcing cannulation were avoided. Further, an appro-priate alternative diagnostic study was elected which permitted confirmation of an obstruction. While outcome was not ultimately satisfactory in this case, the clinical effect of ERCP demonstration of obstruction is of major significance, nonetheless, in diagnosis and treatment.

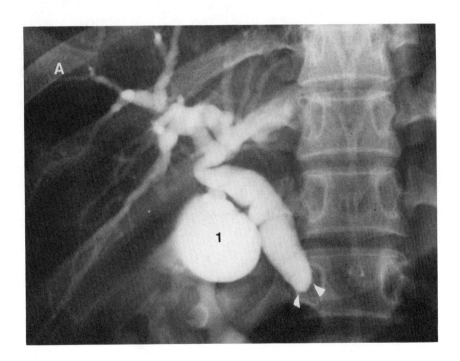

Case 31, Radiograph A

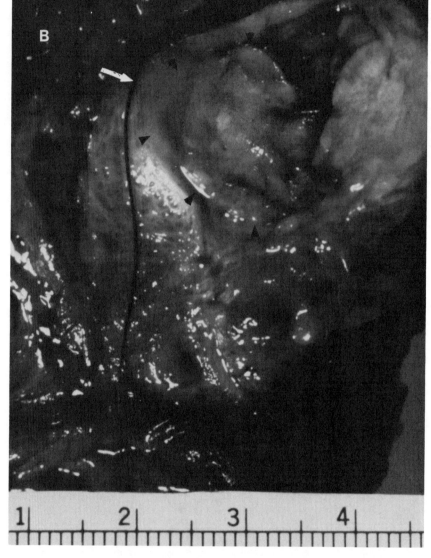

Case 31, Autopsy Specimen B

# NOTES

## CASE 32
### (Accompanying Radiographs A-D)

### Clinical History

A 71-year-old white female presented with a 3-week history of malaise associated with light stool, and concurrent 12-pount weight loss. On examination she was severely jaundiced and hepatomegaly was present. Laboratory studies demonstrated a bilirubin of 19 mg. Upper GI series showed extrinsic pressure on the bulb and liver-spleen scan demonstrated several linear branching defects.

### Radiographic Findings

A. ERCP shows pancreatic duct completely obstructed about 1 cm proximal to the ampulla (arrowheads). This segment of pancreatic duct is curved convexly upward. At the point of obstruction there is a rattail configuration.

B. Percutaneous transhepatic cholangiogram (PTHC), performed because attempted cannulation of the common bile duct was unsuccessful, shows marked dilatation of the common bile duct with complete obstruction distally (arrowheads). The margin of the opaque column at this point is conical with a tendency to beak.

C. RPO and lateral views show gross dilatation of the intrahepatic biliary tree and
& common bile duct. Adequate filling of the distal common bile duct was obtained
D. by placing the patient in the left lateral decubitus and erect standing positions for several minutes.

### Diagnosis

Carcinoma of the head of the pancreas with obstruction of the distal pancreatic and common bile ducts.

### Comment

In contrast to Case 31, the ERCP here demonstrates correlation of the traditional criteria for shape of the occluded segment of the pancreatic duct and carcinoma in that the configuration of the termination is rattail. This traditionally implies the presence of a carcinoma. That information might be sufficient to permit diagnosis except that visualization of the complete pancreatic duct must be achieved or lack of opacification must be accounted for. Thus, in order to confirm that a true obstruction existed (one not due to technical factors), PTHC was performed and confirmed obstruction of the common bile duct. The complementary nature of PTHC and ERCP findings of obstruction at approximately the same level in two separate ducts clearly implies involvement by tumor. This case emphasizes, as others have, that good technique which aims for complete filling of the duct is essential to diagnosis, and here permitted demonstration of the obstruction as a rattail configuration—a feature which is not incidental, but rather achieved through deliberate, difficult procedure, Finally, the extreme dilatation demonstrated in the intrahepatic branches of the biliary system, though not directly pertinent to diagnosis, account for the clinically noted hepatomegaly as well as the linear defects seen on liver-spleen scan.

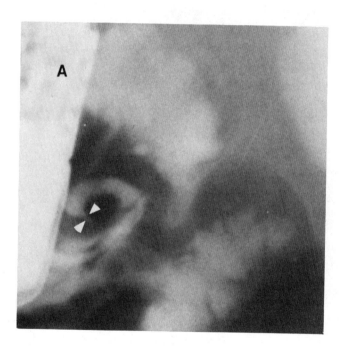

Case 32, Radiograph A

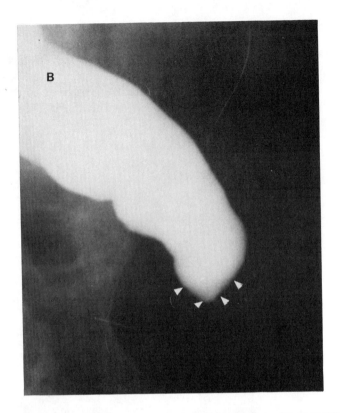

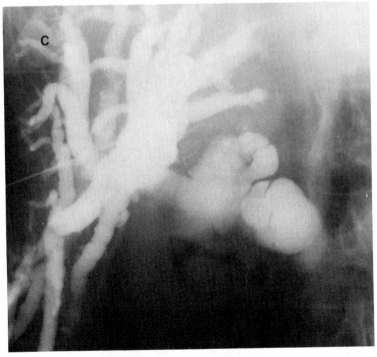

Case 32, Radiographs B & C

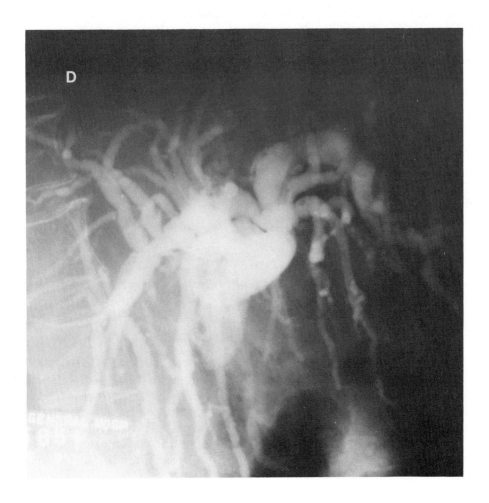

Case 32, Radiograph D

## CASE 33
### (Accompanying Radiographs A & B)

### Clinical History

A 67-year-old black male presented with a history of a 20-pound weight loss over several months with 2 weeks of nausea and pruritus just prior to the current admission. The bilirubin was 10 mg and alkaline phosphatase was markedly elevated. Ultrasound examination showed a mass in the head of the pancreas as well as dilated common bile duct and gallbladder.

### Radiographic Findings

A. There is localized stenosis involving adjacent 1 cm segments of both the pan-
& creatic duct and the common bile duct (arrows) within the head of the pancreas.
B. The stenoses are smooth, fairly abrupt, and proximal to these both ducts are moderately dilated.

### Diagnosis

Carcinoma of the head of the pancreas (confirmed at surgery).

### Comment

The sudden transition from normal to narrowed caliber in this case constitutes an abrupt stenosis typical of that produced by carcinoma (although not exclusively so), whereas a tapering gradual stenosis is seen usually with fibrosis associated with chronic pancreatitis or trauma. That feature notwithstanding, this case warrants the diagnosis of carcinoma because both ducts are adjacently involved (double-duct sign). In our experience this is virtually exclusively due to tumor in the head of the pancreas extending around the adjacent portions of each duct[1,2], although there is recent evidence in the literature that the double-duct sign by itself is not conclusive for a diagnosis of carcinoma (see Case 18, Comment).

### References

1. Freeney PC, Bilbao MK, Katon RM: "Blind" evaluation of endoscopic retrograde cholangio-pancreatography (ERCP) in the diagnosis of pancreatic carcinoma: the "double duct" and other signs. *Radiology* 119:271-274, 1976.

2. Plumley TF, Rohrmann CA, Freeny PC, Silverstein FE, Ball TJ: Double duct sign: reassessed significance in ERCP. *Amer J Roentgenol* 138:31-35, 1982.

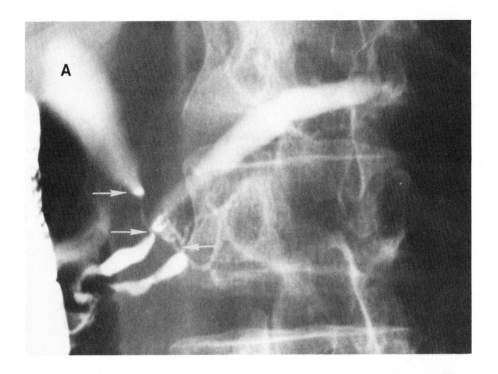

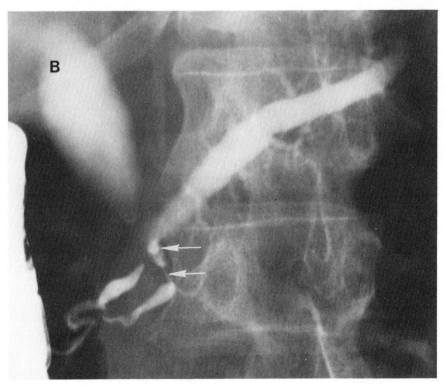

Case 33, Radiographs A & B

**CASE 34**
(Accompanying Radiograph A)

**Clinical History**

A 75-year-old white female was hospitalized for bowel obstruction due to a mesenteric band with arterial compromise and bowel ischemia. At laparotomy the pancreas was noted to be firm in consistency. No biopsy was taken—a provisional diagnosis of chronic pancreatitis was made. ERCP was performed because the patient became jaundiced during the postoperative period.

**Radiographic Findings**

A. The pancreatic duct is occluded within the head about 3 cm proximal to the ampulla (arrow). The termination of the occluded segment is conical and has irregular margins. The common bile duct shows severe, although smooth, tapering stenosis about 1 cm cephalad to this point, and is normal in caliber distally (arrowhead).

**Diagnosis**

Carcinoma of the head of the pancreas (confirmed at surgery).

**Comment**

As with Case 33, we see here double-duct involvement, although the pancreatic duct is occluded in this case rather than stenosed. The irregular margins at the point of stenosis would themselves be an indication of malignancy. As noted in Case 30, the configuration of the stenosis or occluded segment of the pancreatic duct or common bile duct is secondary in significance, diagnostically, to the fact that stenosis is present. Here, even that feature is secondary to the fact that both ducts are involved at their adjacent points with stenosis, implicitly suggestive of the presence of carcinoma as discussed in Case 33.

The clinical history of palpation of the pancreas, as well as the resulting clinical impression of chronic pancreatitis, has not been given here for the purpose of imputing error, but to demonstrate, on the contrary, that very careful examination was made. Thus, if the finding is in error, it is one inherent in diagnostic dependency on that technique itself: without accompanying biopsy, palpation alone is simply nondiagnostic.

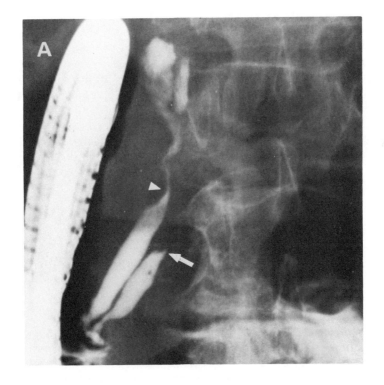

Case 34, Radiograph A

## CASE 35
### (Accompanying Radiographs A & B)

### Clinical History

A 72-year-old black female presented with a 6-week history of obstructive jaundice, right upper quadrant tenderness, and hepatomegaly on examination. The bilirubin was 10 mg and a liver-spleen scan demonstrated defects consistent with metastatic tumor.

### Radiographic Findings

A. ERCP shows filling with slight overdistention of the distal pancreatic duct (1) and several side branches in the head. There is a long, smooth, tapered stenosis (arrows) proximal to this region.

B. Proximal to the stenosis, extravasation of the contrast material in an amorphous fluffy collection (arrowheads) is noted. There is no filling of the pancreatic duct proximal to this point.

### Diagnosis

Carcinoma of the head of the pancreas (confirmed at surgery).

### Comment

Stenosis, by itself, especially when it involves a long segment, may be due to benign stricture as often as to tumor. However, demonstration of extravasation of the contrast material at the site of stenosis is characteristic of and unique to tumor. Usually, it appears to be due to extravasation of the contrast material directly from the lumen of the pancreatic duct into interstices of the tumor mass. Extravasation resulting from duct trauma during the procedure is not appropriately part of the differential in any event; in addition, this has not occurred in our 5-year experience (see also Case 36).

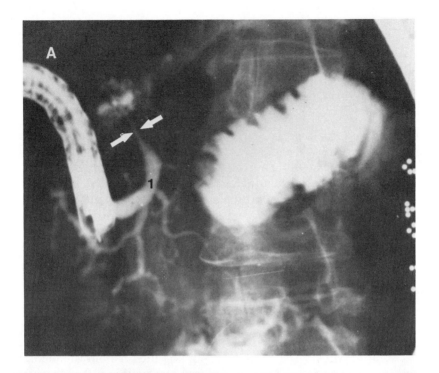

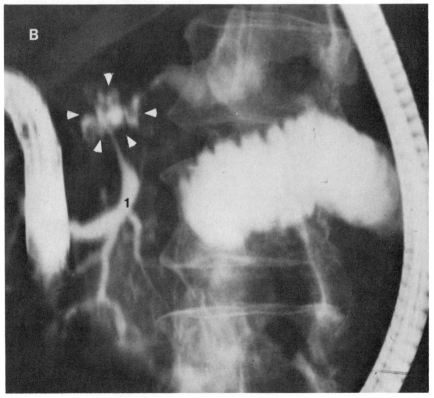

Case 35, Radiographs A & B

## CASE 36
### (Accompanying Radiographs A & B)

### Clinical History

A 63-year-old white female presented with a 1-week history of pruritus and midepigastric pain radiating to the back. On admission, she was jaundiced and had right upper quadrant tenderness. Bilirubin was 11 mg and alkaline phosphatase was moderately elevated. Past medical history included a hysterectomy for a benign condition about 2 years prior to the current admission. At that time she was noted to have gallbladder calculi, although cholecystectomy was not performed.

### Radiographic Findings

A. Upper GI series shows abnormal mucosa on the inner aspect of the second part of the duodenum with mucosal spiking and edema (arrows).

B. ERCP shows the endoscopy catheter (arrow) in the distal pancreatic duct (1) which is stenotic within the head. Surrounding the segment of stenosis are irregular spidery collections of contrast material (arrowheads) due to extravasation. No filling of the pancreatic duct proximal to this region could be achieved, nor could the common bile duct be cannulated.

### Diagnosis

Carcinoma of the head of the pancreas.

### Comment

Subsequent surgery revealed a large tumor mass near the distal common bile duct accounting for the failure of cannulation. See discussions for Cases 31 and 32 with regard to the significance of failure of cannulation. This case presents an example of extravasation associated with stenosis, which, as noted in Case 35, is usually due to carcinoma. Mucosal changes on the inner adjacent aspect of the C loop may be associated with tumor or inflammatory processes, especially in the head of the pancreas. Thus, cholecystitis may produce such dramatic changes, although rarely, and even then they usually affect the lateral aspect. Therefore, the edema and spiking here are directly related to tumor.

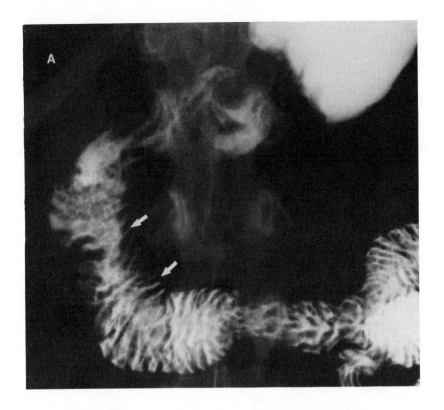

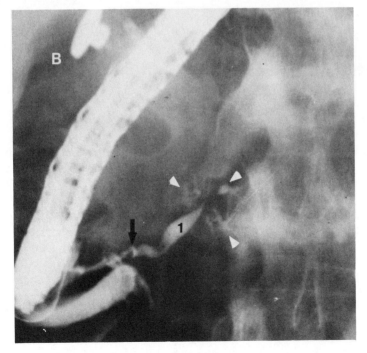

Case 36, Radiographs A & B

CASE 37
(Accompanying Radiographs A-C)

## Clinical History

A 38-year-old black female presented with a 6-month history of weight loss and back pain. On examination at the current admission, a 12 cm round, hard, epigastric mass was discovered. Oral cholecystography showed multiple, small, radiolucent calculi.

## Radiographic Findings

A. ERCP shows a normal, moderately overdistended pancreatic duct with filling of
&  side branches and acinarization in the head. There is stenosis commencing about
B. 3 cm proximal to the ampulla and extending throughout the body (arrowheads). In this region, the duct is approximately 1 mm in diameter and is uniformly smooth.

C. Celiac arteriogram shows filling of the major visceral arteries. There is encasement of the celiac, hepatic, left gastric, and splenic arteries for several centimeters at their proximal distribution with lumen reduction to about one-half normal.

## Diagnosis

Carcinoma of the body of the pancreas.

## Comment

Long-segment stenosis (here 8 cm) without any features associated with chronic pancreatitis such as irregular margins, dilatation, or side-branch abnormalities, indicates the presence of tumor, particularly since the stenosis seen is fixed, with no change in caliber as a variable of the pressure of injection from Radiograph A to B. Encasement of the pancreatic duct produced by the large tumor mass produces the regular 1 mm diameter fixed contrast material column which represents the true caliber of the narrowed pancreatic duct. (Acinarization seen in the head is an indication of the thoroughness of examination.) Here, the arteriogram confirms ERCP findings of extensive tumor by demonstrating encasement of multiple large branches of the celiac trunk.

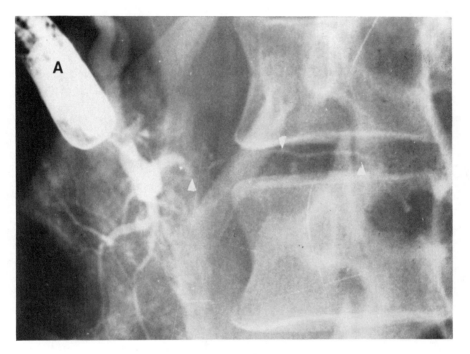

Case 37, Radiograph A

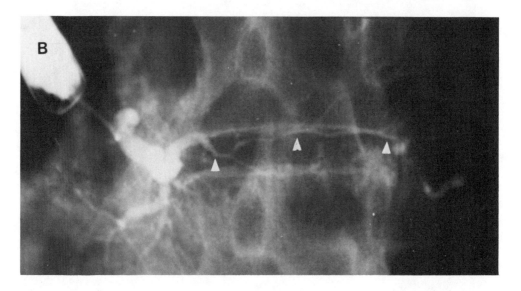

**Case 37, Radiograph B**

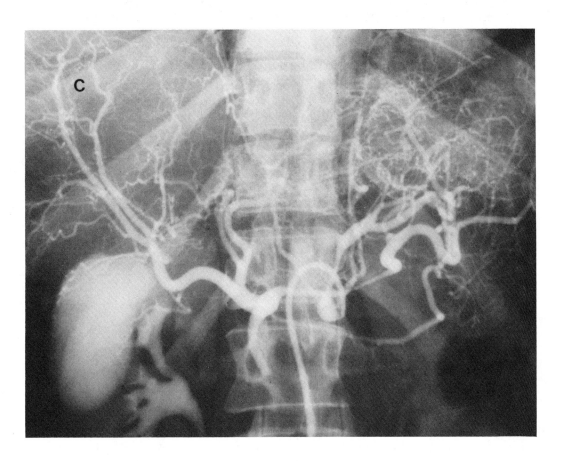

Case 37, Radiograph C

**CASE 38**
(Accompanying Radiographs A-C)

## Clinical History

A 72-year-old white male presented with a 6-month history of abdominal pain. A duodenal ulcer was previously diagnosed by upper GI series and had responded only minimally to antacid treatment. About 1 month prior to current admission he had developed back pain which increased in severity. Upper GI series on this admission showed a small ulcer crater in the duodenal bulb, confirmed at endoscopy, and the C loop was normal. ERCP was performed because of the change in character of the pain and its nonresponsiveness to antiulcer regimen.

## Radiographic Findings

A. ERCP, AP and LPO views, show the pancreatic duct in its entirety. There is
B. moderate overfilling with side-branch visualization in the proximal body and tail
& in addition to some acinarization. Despite this deliberate effort at complete
C. filling, a persistently narrowed segment about 2 cm in length is present in the head (arrows). This has smooth margins and is not associated with caliber variation elsewhere.

## Diagnosis

Carcinoma of the head of the pancreas (confirmed at surgery).

## Comment

An otherwise featureless localized stenosis would usually not permit a definitive diagnosis. However, in this case, given the absence of radiological and clinical evidence of pancreatitis, along with the well-defined clinical setting (especially persistent back pain), the differential is narrowed to the strong probability of carcinoma. A single localized segment of stenosis in chronic pancreatitis must be regarded as rare in any event.

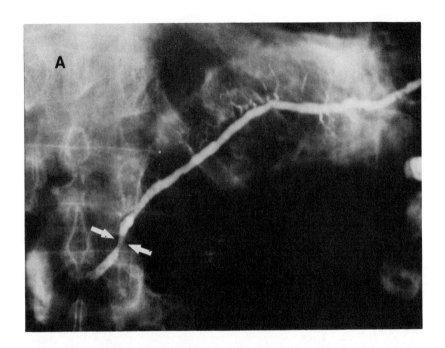

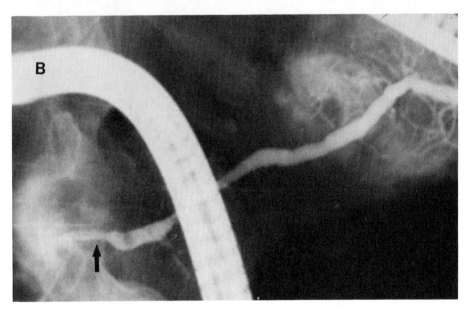

Case 38, Radiographs A & B

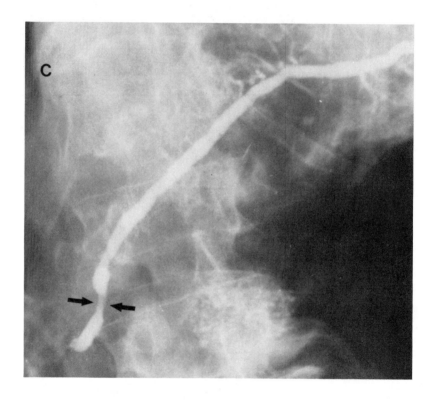

Case 38, Radiograph C

# VI

# Self-Evaluation

# INTRODUCTION

This self-evaluation section intends to test, more than diagnostic capability, attention to assessment processes. Thus, while diagnosis is withheld, the main focus of self-evaluation is not whether one's diagnosis matched the case, but rather whether all details in the radiographic descriptions and commentaries were considered. It seems useful in interpreting the cases which follow, to focus on the similarities of pattern and distinctive relationships as they present likeness and contrast to cases in Sections II through V.

Each case is presented with such clinical detail as is necessary and a radiograph(s) which is considered conclusive in itself for diagnosis. On the page following these items are the remaining elements: Radiographs, Descriptions, Diagnosis, and Comment. It is suggested that the reader, prior to turning to the "answer," write his own observations in the order noted above for comparison.

## CASE 39
### (Accompanying Radiographs A-F)

**Clinical History**

A series of six separate cases. ERCP was performed in each for exclusion of pancreatitis.

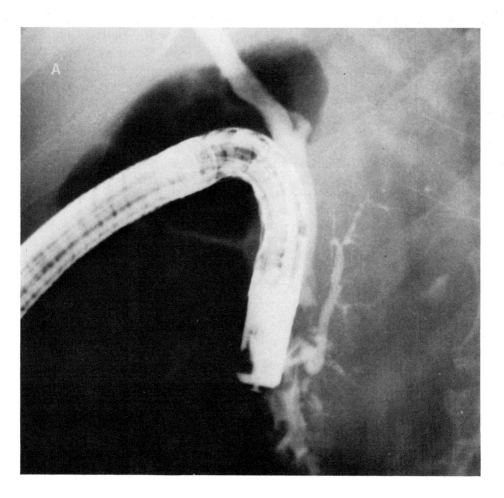

**Case 39, Radiograph A**

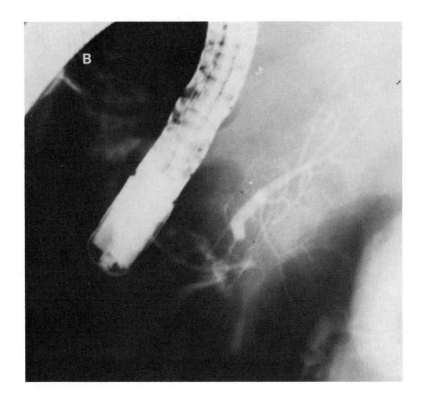

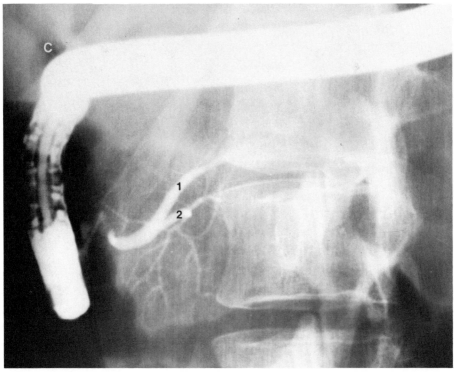

Case 39, Radiographs B & C

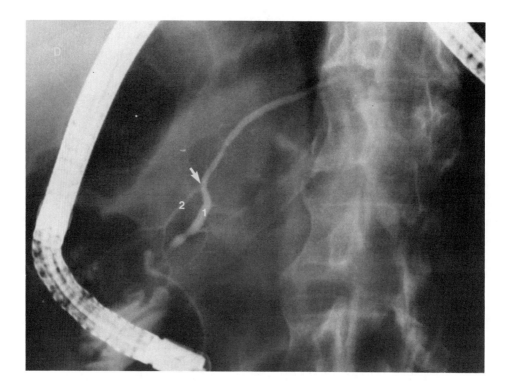

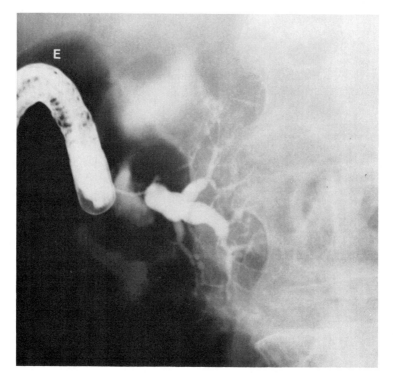

Case 39, Radiographs D & E

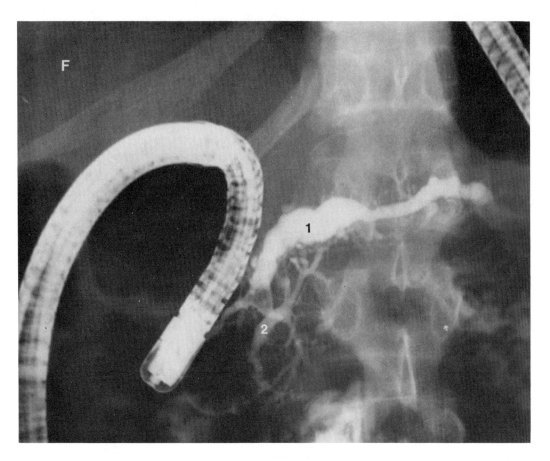

Case 39, Radiograph F

**Radiographic Findings**

A. The accessory pancreatic duct is opacified and progressive, regular, smooth tapering is noted, commencing from its termination at the duodenum. There is no filling of the main pancreatic duct and no duct structures in the body and tail are shown.

B. There is somewhat more filling of the accessory duct alone than occurred in A. Branches throughout the head of the pancreas are noted. There is no visualization of the main pancreatic duct.

C. There is partial filling of the main pancreatic duct (1) as well as a large accessory duct (2), supplying the head both medially and laterally to the main duct.

D. A large accessory duct (2) coursing almost entirely along the right lateral aspect of the head is visualized. Its origin from the main pancreatic duct (1) from the apex of its cephalic curve (arrow) is a common configuration.

E. Catheterization of the accessory pancreatic duct shows significant overdistention resulting from the attempt to obtain proximal filling and visualization of the main pancreatic duct. Injection was discontinued on appreciation that the accessory duct was independent of the main pancreatic duct.

F. The ERCP catheter lies within a large accessory duct (2). Extensive filling of the main pancreatic duct (1) occurs through communication between the two systems. The main pancreatic duct itself does not appear to communicate directly with the duodenum. The appellation "pancreatica divisa" has been given to this appearance.

**Diagnosis**

In these cases, the anatomic variant is the focus of the self-evaluation. Diagnosis is essentially not at issue here.

**Comment**

In the vast majority of cases the duct of Wirsung serves as the main pancreatic duct. The duct of Santorini may be represented by a few side branches filling in the head of the pancreas or may open to the duodenum through an accessory papilla, proximal to the main pancreatic duct.

In about 1% of cases, the main pancreatic duct (Wirsung) is small, and the principal drainage for the pancreas is through a large duct of Santorini. Even more rarely, the duct of Wirsung is small and isolated, so that injection shows filling only of duct structures within the head and no filling of body and tail is obtained.

The clinical importance of recognition of the accessory duct lies in the distinction from an erroneous diagnosis of main duct obstruction. Example F results from malfusion of the ventral and dorsal embryological elements of the pancreas. The extent of malfusion varies from a dominant Wirsung situation, through a dominant Santorini system, to a completely isolated small duct of Wirsung.

Even when there is a separate opening for an accessory duct, either directly to the duodenum or through a shared papilla, cannulation may be difficult and thus the portion of the pancreas served by the accessory duct may not be visualized.

## CASE 40
### (Accompanying Radiographs A-C)

**Clinical History**

A 48-year-old female with a 2-year history of recurrent abdominal pain and abnormal liver function studies, but no jaundice.

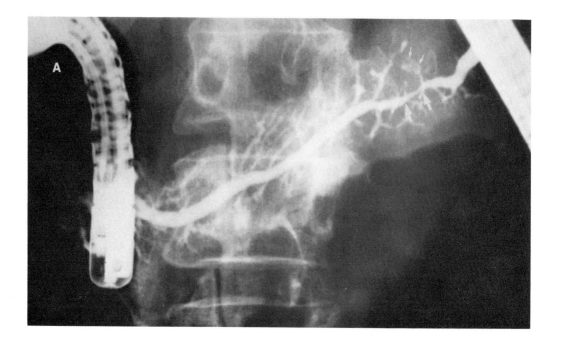

**Case 40, Radiograph A**

# NOTES

## Radiographic Findings

A. ERCP shows the pancreatic duct in its entirety (the tail, not illustrated, was clearly visualized on other views). A short-segment stenosis is seen in the proximal third (arrows) which shows a smooth, tapered configuration and is associated with side-branch irregularity as well as a small aneurysm (small arrows). Prominent side-branch filling is seen throughout the body and head, and the general caliber of the duct is at the upper limits of normal. This appearance is secondary to intentional, controlled overfilling in an attempt to distend the narrowed segment.

## Diagnosis

Chronic pancreatitis with localized stricture.

## Comment

As with Case 38, a localized stenosis without further characterizing features usually does not permit definitive diagnosis. However, in this case distinctive features of chronic pancreatitis are present in the area immediately adjacent to the stricture: side-branch irregularity, small duct aneurysm. Note that, occasionally, deliberate overfilling is employed to distinguish between true and only apparent stenosis.

Two additional cases of chronic pancreatitis are presented to emphasize the varying presentations of stenosis.

B. This is another patient with moderately severe chronic pancreatitis (46-year-old female with a 3-year history of nausea, vomiting, and abdominal pain). Here is the typical appearance of chronic pancreatitis: multiple areas of stenosis alternating with uninvolved, normal caliber duct. Minimal irregularity of side-branch architecture is also present. Although some of the stenoses are quite severe, the general pattern is of diffuse involvement (see also Cases 15 and 16).

C. A 48-year-old male with a long history of recurrent pancreatitis. The case is shown for the demonstration of long-segment stenosis in the body and head (arrows), showing a smooth, tapered appearance both proximally and distally. This form is characteristic of simple stricture, confirmed in this case by changes of chronic pancreatitis both within the head, body, and tail, where duct dilatation and aneurysms are present. Compare this with Cases 17 and 37, each of which show single long-segment stenoses throughout the body and head, with relatively smooth tapered transition. Both cases demonstrate that the presence or absence of distinguishing features of chronic pancreatitis permits the diagnosis of simple stricture (Cases 40 C and 17) versus carcinoma (Case 37).

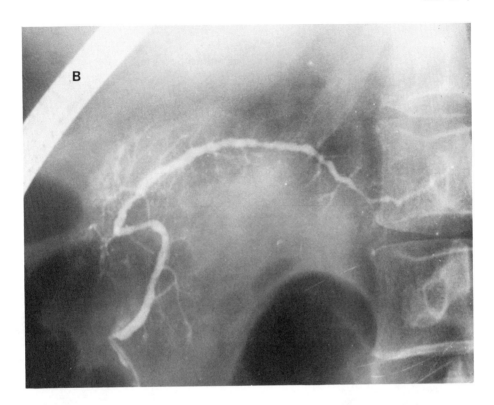

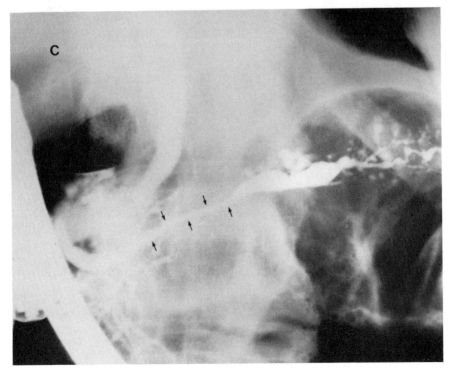

Case 40, Radiographs B & C

## CASE 41
### (Accompanying Radiographs A & B)

**Clinical History**

A 60-year-old black female with a 55-pound weight loss over the previous 12 months. Her appetite was diminished accompanied by anorexia and nausea. She had chronic abdominal pain. Liver function studies were mildly abnormal.

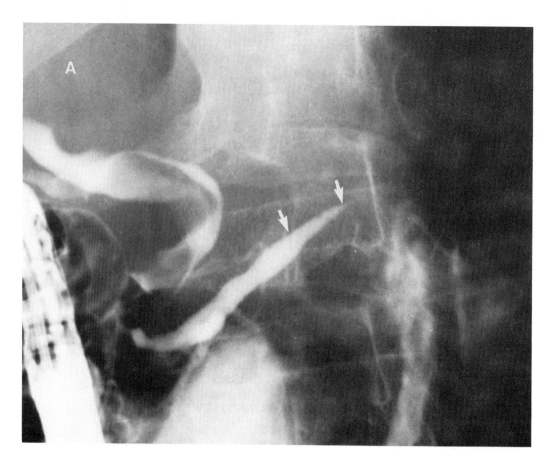

Case 41, Radiograph A

# NOTES

## Radiographic Findings

A. ERCP shows a smooth tapered stenosis (arrows) at the middle third of the pancreatic duct with moderate duct dilatation up to this point. There is complete occlusion with no proximal filling. The stenosis extends over a segment approximately 1 cm in length, and shows a smooth tapered rattail-shaped termination. This feature alone should immediately suggest the diagnosis of carcinoma (which in this case was confirmed at laparotomy, where a mass was palpable in the tail of the pancreas, and metastatic tumor was noted in the liver). See Case 35 which demonstrates a similar stenosis.

## Diagnosis

Carcinoma of the pancreas (body).

B. ERCP performed on another patient in a similar clinical setting to rule out carcinoma of the pancreas. The pancreatic duct is occluded in its middle third. Up to this point it is distended and the side branches show overfilling in that too many are visualized and appear larger in caliber than is usual with normal structure. The stenosis is characterized for involving a short segment with a smooth, somewhat beak-shaped termination (arrows). Note that there is no filling of the duct beyond this point. Compare Case 34.

## Diagnosis

Carcinoma of the pancreas (body).

## Comment

Although the precise configuration may vary somewhat, the significant feature here is complete occlusion of the pancreatic duct. Total obstruction is always strongly suggestive for carcinoma. Cases 21, 23, and 24, for example, show severe stenosis of the pancreatic duct in association with pseudocyst or chronic pancreatitis with proximal duct dilatation. Yet, with the most severe of stenoses, some proximal duct filling is possible. Thus complete duct obstruction is significant for, if not in fact singularly diagnostic of, neoplasm. Although calculi may theoretically completely occlude the pancreatic duct, we have not observed this to occur in practice.

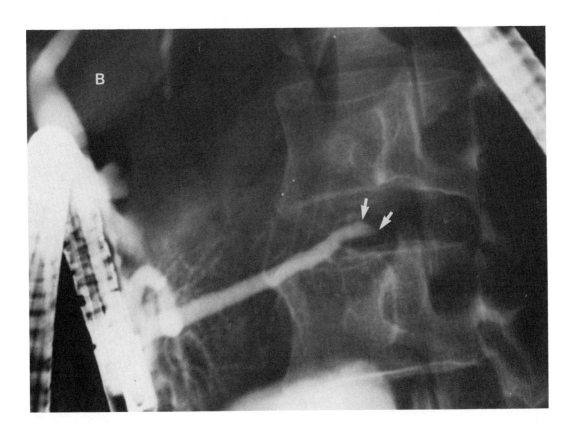

Case 41, Radiograph B

# CASE 42
## (Accompanying Radiographs A-C)

**Clinical History**

A 36-year-old male admitted for evaluation of a mass in the left upper quadrant. He had a history of several weeks duration of nonspecific abdominal pain associated with vomiting. He denied excessive alcohol intake or recent trauma.

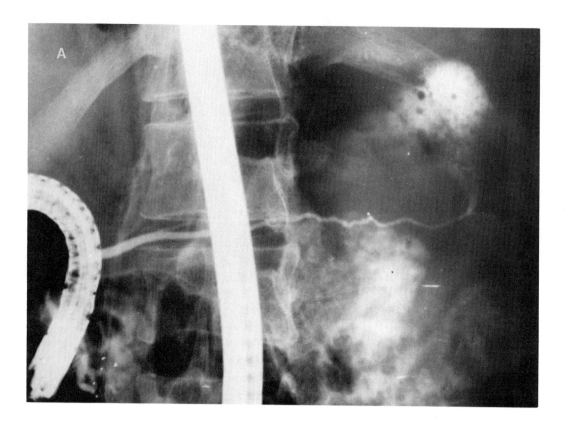

Case 42, Radiograph A

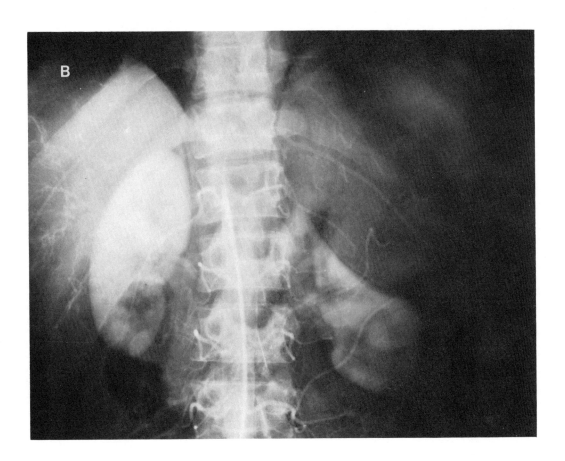

Case 42, Radiograph B

## Radiographic Findings

**A.** ERCP shows filling of the entire pancreatic duct, which is normal in caliber and shows no abnormality except in the tail, where there is free communication between the pancreatic duct and an irregular cavity in the left upper quadrant. The pancreatic duct is smooth and shows no localized narrowing or dilatation at the site of communication, nor evidence of any filling defect. The cavity outlined is featureless and does not correspond to any recognizable anatomic structure. Filling defects within it are presumably due to debris.

**B.** This view of the abdomen, taken at a late phase during aortography, shows deformity and displacement of the left kidney by a large rounded structure which almost entirely fills the left upper quadrant. This shows smooth contours which are well-defined.

## Diagnosis

Pancreatic pseudocyst (tail).

## Comment

This example recalls Case 24 where direct communication between the pancreatic duct system and a pseudocyst was also present. ERCP is not used as the primary diagnostic procedure for pseudocyst and, in fact, is usually not employed at all as the diagnosis generally is established by ultrasound. ERCP may be performed when the pseudocyst is complicated as in this case, or when there is suspicion of duct abnormality, especially in the face of clinical nonresolution of the pseudocyst.

It is somewhat surprising that the pancreatic duct is entirely normal in this case. This unusual feature suggests that the process which caused the pseudocyst is confined exclusively to the tail of the pancreas.

As mentioned previously in relation to Cases 24 and 25, intraoperative pancreatic cystography (Radiograph C) may assist in determining the size of the structure (arrowheads) and direction of tracking. While such esoteric studies may aid in directing the surgical approach and choice of drainage procedure, this is highly individual with the operator—and possibly a rare need. Radiograph A demonstrates the extent of opacification recommended for pseudocyst. Sufficient contrast material injection for comprehensive outline of the structure should not be done on ERCP because it is superfluous to therapy and in general the more contrast material injected, the higher the incidence of complication.

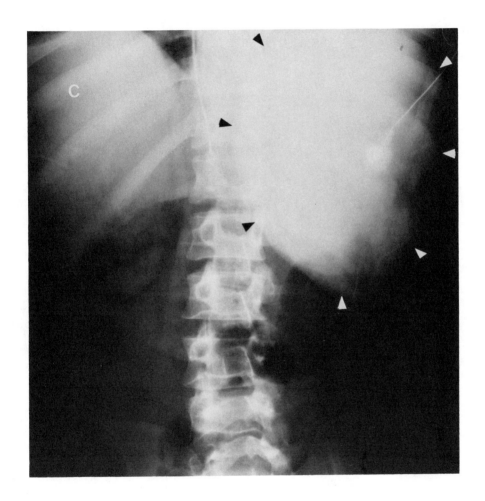

**Case 42, Radiograph C**

# CASE 43
## (Accompanying Radiographs A-G)

## Clinical History

A 3-year-old female with a 5-month history of vomiting, abdominal pain, and jaundice. Her mother noticed yellowing of her eyes and darkening of her urine. Serum bilirubin varied between 2 and 8 mg/100 ml.

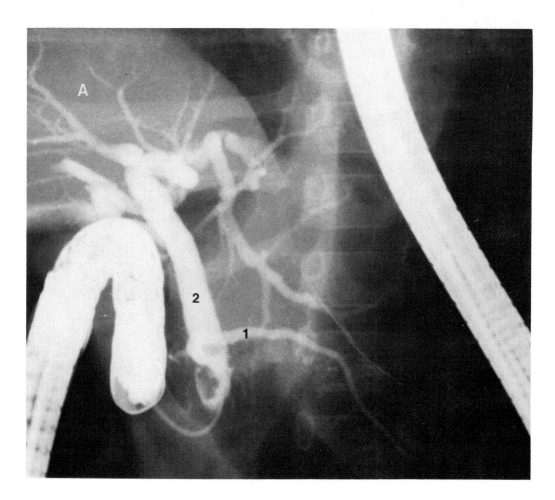

**Case 43, Radiograph A**

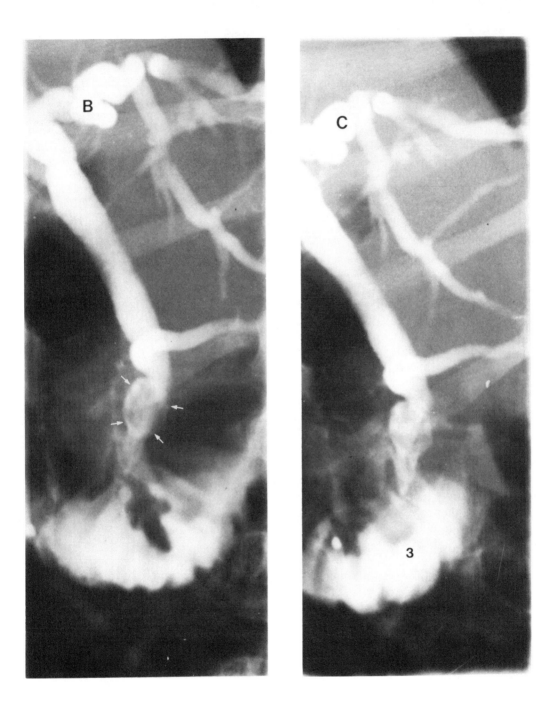

Case 43, Radiographs B & C

## Radiographic Findings

A. ERCP shows filling of both the pancreatic duct (1) and common bile duct (2). A persistent filling defect approximately 1 cm in diameter is located within the distal aspect of the common duct just proximal to the ampulla, and contrast material passes in a narrow layer between the defect and the margin of the common bile duct. The latter is moderately dilated.

B. Detail views of the distal common bile duct show the irregular filling defect
& (arrows) more clearly. Contrast material outlines the second portion of the
C. duodenum (3).

## Diagnosis

Choledochocele with obstructive jaundice.

## Comment

After laparotomy and duodenostomy, a small protuberant mass was noted in the medial aspect of the second portion of the duodenum, in the area of the ampulla. Sphincterotomy was performed and resulted in visualization of the neck of a choledochocele, which communicated with the distal common bile duct. The cyst was unroofed with relief of symptoms.

The usual location for a choledochocele is in the common bile duct, between the ampulla and the cystic duct entry. It may range in size from a small lesion such as shown here, to large cysts that involve the entire common bile duct.

The chief diagnostic feature here is intralumenal location of the lesion. Diagnosis is achieved through recognition of the smooth contour of contrast material within the distal common bile duct, and by appreciation, as well, of a thin layer of contrast material around the filling defect on all sides. It is essential to perform the examination in multiple projections.

A brief summary of three additional cases follows for their interest as they reflect other kinds of common bile duct problems.

D. ERCP in this 78-year-old male shows moderate common bile duct dilatation (1), without localized obstruction. A localized collection of contrast material (arrows) overlies the distal common duct. This communicates with the duodenum in an area that shows mucosal folds.

## Diagnosis

Common bile duct dilatation, moderate, cause unknown. Duodenal diverticulum.

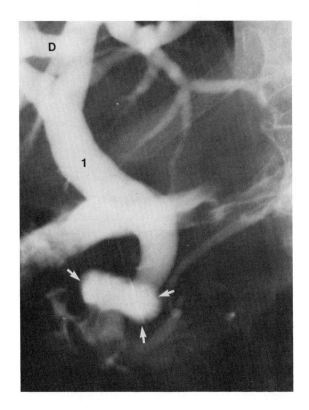

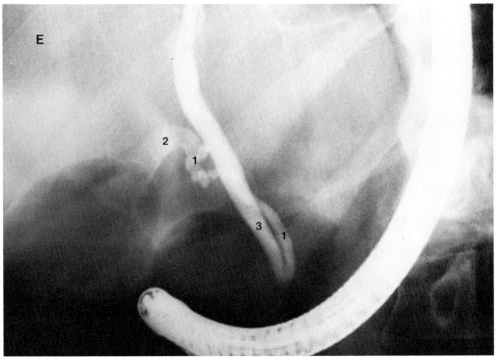

Case 43, Radiographs D & E

## Comment

It is tempting to relate the findings of this case on a causal basis. However, duodenal diverticulum is a common condition, and in the vast majority of cases causes no obstruction of the biliary system. The diverticulum is lined by mucosa and communicates directly with the second portion of the duodenum. Usually, diverticula are no larger than 2 cm in diameter. Deformity of the biliary system and distal pancreatic duct might result from very large diverticula, but these are quite uncommon. Of more pragmatic concern, since the diverticula are often periampullary in location, is their potential for interfering with cannulation of the ampulla.

E.  A 40-year-old male with a normal common bile duct and intrahepatic ducts, studied for unexplained abdominal pain. The noteworthy feature is the location of the cystic duct (1), which passes from the gallbladder infundibulum (2) and courses medially and caudally, to join the distal common bile duct (3) almost at the ampulla. The result of this arrangement is that cystic and common bile ducts run approximately parallel for much of their course. This arrangement is of surgical importance in that its recognition is not possible by palpation or direct visual inspection, and ERCP definition avoids dissection or other procedures which might place the common bile duct at risk.

F.  ERCP in a 35-year-old male, admitted with a 2-week history of malaise and
&  abdominal pain. Examination showed a normal pancreatic duct and intrahepatic
G.  radicals (not illustrated). The common duct is normal in overall caliber, shows a persistent linear radiolucent defect (arrows) extending at least from the cystic duct level to the ampulla. Cystic duct and gallbladder are clearly visualized and are normal. The linear filling defect is 1 to 2 mm wide, shows a smooth defined margin on each side, and has an almost "mechanical" regularity and smoothness.

## Diagnosis

E.  Anomalous medial insertion of the cystic duct.
F. & G.  Anomalous low insertion of the cystic duct.

## Comment

These cases represent variants of the normal pattern of cystic duct insertion. Other causes of persistent linear filling defects in the common duct include parasitic infestation, principally ascaris, although these are usually larger in caliber and frequently occlude the common duct[1-3]; unusual linear mucosal fold; or retained catheter fragment.

The conditions shown here which affect the distal common bile duct illustrate the variety of near-normal states that mimic disease. Not all filling defects are due to calculus or tumor; not all abnormalities are due to disease.

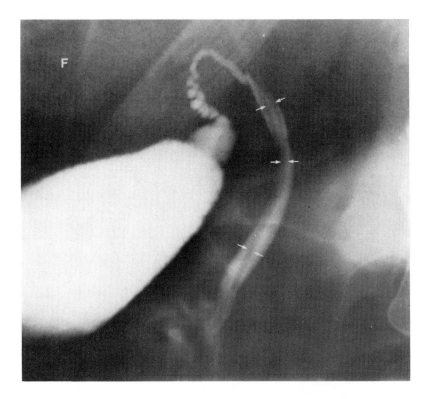

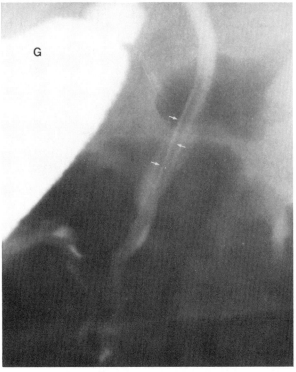

Case 43, Radiographs F & G

## References

1. Cremin BJ, Fisher RM: Biliary ascariasis in children. *Amer J Roent* 126:352-357, 1976.

2. Louw JH: Abdominal complications of ascaris lumbricoides infestation in children. *Brit J Surg* 53:510-521, 1966.

3. Wright RM, Dorrough RL, Ditmore HB: Ascariasis of the biliary system. *Arch Surg* 86:402-405, 1963.

# NOTES

# CASE 44
## (Accompanying Radiographs A-G)

### Clinical History

A 64-year-old male with a 4-month history of malaise, weakness, dark urine, and yellow skin pigmentation. Jaundice has persisted following cholecystectomy for cholelithiasis 2 months previously, and ERCP was performed to rule out common duct calculus.

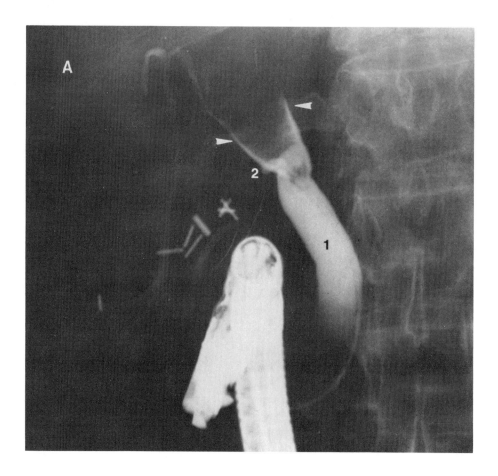

**Case 44, Radiograph A**

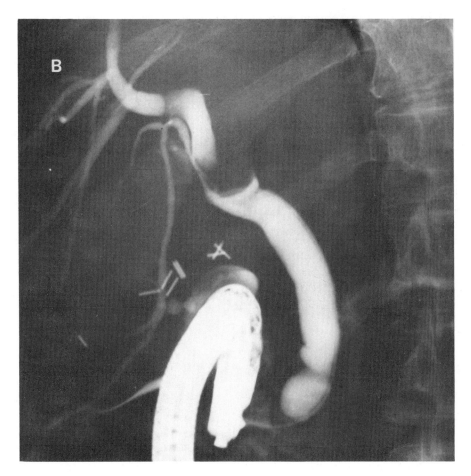

Case 44, Radiograph B

## Radiographic Findings

A. The catheter has been passed proximally up the common bile duct (1) and the tip is located adjacent to a persistent filling defect within the duct at the expected level of the cystic duct (2). This shows a convex inferior margin. Its upper aspect is not defined. A thin layer of contrast material outlines the lateral margins of the filling defect (arrowheads).

B. With injection of additional contrast material slight proximal filling of intrahepatic structures occurs and these ducts appear normal.

## Comment

Smooth filling defects are typical for calculus. When the defect is facetted or mobile there is essentially no differential for the diagnosis. When the filling defect is completely outlined by contrast material on all sides, diagnosis is relatively unequivocal. However, when the appearance of the contrast material column is meniscus-shaped, as it is here, one infers that the column is in contact with a portion of a calculus.

The following cases are representative of the range of irregular and smooth stenoses of the common duct.

C. ERCP performed in a 70-year-old female with symptoms of obstructive jaundice. There is a persistent filling defect in the common hepatic duct (1) which is partially outlined by contrast material around it. There is persistent narrowing of the left main hepatic duct (arrow) with proximal dilatation.

## Diagnosis

Carcinoma of the common hepatic duct.

## Comment

As with Case 44 A and B, the features are those of an intrinsic lesion, somewhat less well defined. The presence of localized stenosis in the left main hepatic duct suggests the presence of malignancy. The chief diagnostic difficulty here lies in distinguishing intrinsic from extrinsic processes. Pathologically, when the biliary system is invaded by tumor, this may appear as an *intrinsic* abnormality, especially where submucosal nodular lesions result. The distinction, diagnostically, is made on the basis of associated features, such as mass effect, and at times is not possible.

Over one-half of the primary tumors of the bile ducts occur in an area including the common hepatic, the left or right hepatic ducts, or their confluence. The eponym, Klatskin tumor[1,2], is applied when this involvement occurs at or near the bifurcation of the common hepatic duct and is due to adenocarcinoma, which is by far the most predominant cell type.

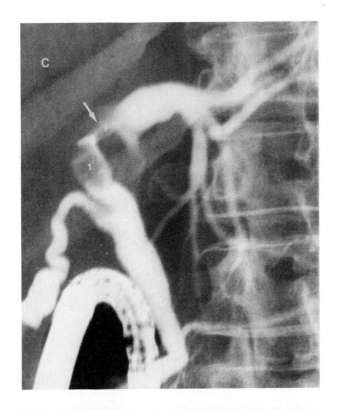

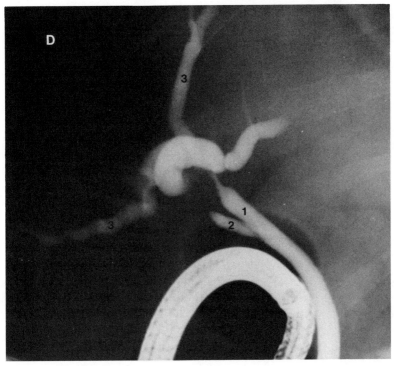

Case 44, Radiographs C & D

D. ERCP performed in a 59-year-old female with symptoms of mild obstructive jaundice. Cholecystectomy for cholelithiasis was performed many years previously. The common bile duct (1) and cystic duct stump (2) are visualized. Intrahepatic radicals are identified (3) and are moderately dilated. There is a localized smooth tapered stenosis of the common hepatic duct which extends proximally to the junction of the main right and left hepatic ducts. No filling defects are associated.

## Diagnosis

Benign stricture of the common hepatic duct, postsurgical.

## Comment

Postoperative strictures of the extrahepatic biliary system typically produce a simple stenosis; that is, localized smooth, tapered narrowing consistent with benign processes, typically, given a supportive clinical history.

E. A 55-year-old white male who presented with chills, fever, and jaundice. He had
& sustained common bile duct injury at cholecystectomy 15 years prior to this
F. admission and had had multiple attempted repairs since. Bilirubin 15 mg/100 ml retrograde cholangiography (Radiograph E) shows a long segment stenosis (arrows) involving the common bile duct and common hepatic duct, and extending to the junction of the hepatic ducts at the hilum of the liver. The maximum caliber of the stenosis is about 2 mm. The entire biliary system is markedly dilated up to the site of stenosis. Intraoperative transhepatic cholangiogram (Radiograph F) shows the stenotic segment more clearly (arrows).

## Comment

Common bile duct injury consequent to previous surgery may be associated with distorted anatomy and scarring. In this and the preceding patient (Case 44 D), demonstration of the cephalad extent of the stenosis provides essential information regarding, first, the feasibility of reconstructive surgery and, secondly, the choice of procedure and approach used. Thus, precise delineation of anatomy is integral to any attempt at surgical reconstruction.

G. Retrograde cholangiogram performed on a 59-year-old male with a 1-year history of chills, 50-pound weight loss, and anorexia. There is marked dilatation of the common bile duct (1) and of the cystic duct remnant (2). There is very severe, localized stenosis of the distal common duct, where the lumen is reduced to thread-like caliber (white arrows). The margins of the stenotic area form an acute angle and appear lobulated proximally (open arrow) where they create the angle in relation to the common duct margin (black arrows).

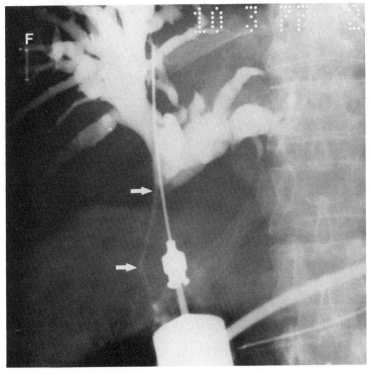

Case 44, Radiographs E & F

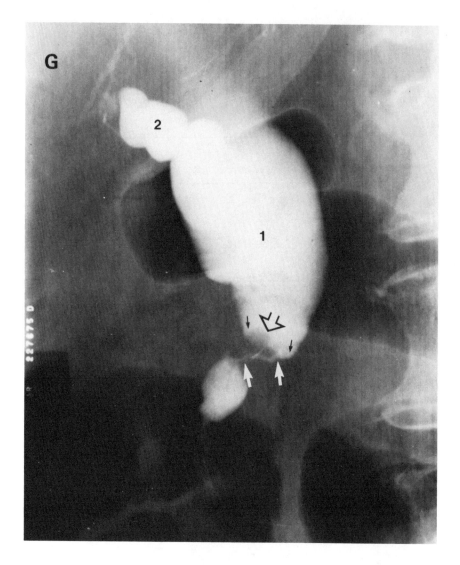

Case 44, Radiograph G

## Diagnosis

Carcinoma of the common bile duct with severe proximal obstruction.


## Comment

This lesion has all of the classic features of an intrinsic malignant process. Ultrasound examination confirmed marked dilatation of intrahepatic and extrahepatic biliary system, and also showed an echogenic lesion within the liver, presumably metastatic tumor. We include this example as a typical case of primary intrinsic malignant involvement of the common bile duct. It shows the typical acute angle produced between the tumor mass and the common duct margin, which is analogous to the "apple core" radiologic appearance encountered with neoplasms in other organs and defines an intrinsic process (see Cases 31-34 for examples of extrinsic tumor involvement of the common duct).


## References

1.  Klatskin G: Adenocarcinoma of the hepatic duct at its bifurcation within the porta hepatis. *Am J Med* 38:241, 1965.

2.  Longmire WP, McArthur MS, Bastounis EA, Hiatt J: Carcinoma of the extrahepatic biliary tract. *Ann Surg* 178:333-343, 1973.

**CASE 45**
(Accompanying Radiograph A)

## Clinical History

This 82-year-old woman with chronic lymphocytic leukemia had had a cholecystec-tomy approximately 2 weeks prior to this examination. She had pulled the T tube out a few days after surgery, as a result of which a cutaneous and intraperitoneal fistula developed.

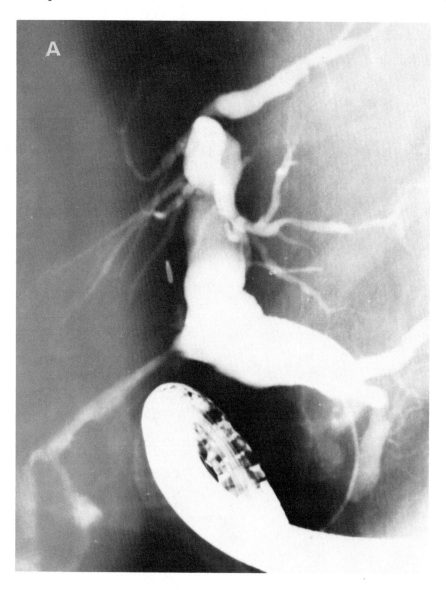

Case 45, Radiograph A

# NOTES

## Radiographic Findings

**A.** Retrograde cholangiogram shows a moderately dilated common bile duct with filling of several intrahepatic branch ducts which are normal in caliber. A linear collection of contrast material fills from the lateral aspect of the common bile duct, at approximately the level of the cystic duct.

## Diagnosis

Choledochocutaneous fistula secondary to self-removal of T tube.

## Comment

The current study shows a single fistula along the course of the T tube. Despite the difficulty of endoscopy in the early postoperative phase, its use may obviate re-exploration in a patient whose condition is already compromised. The present study demonstrated that forceful self-removal of the T tube had not significantly traumatized the common duct.

**NOTES**

## CASE 46
### (Accompanying Radiographs A-D)

**Clinical History**

A 37-year-old female enjoyed good health until 2 years previously, when she developed lassitude and generalized pruritus. Liver function studies at that time were abnormal and total bilirubin was increased. On surgical exploration the gallbladder contained many bilirubin calculi. Cholecystectomy and choledochoduodenostomy were performed. From that time her course was marked by recurrent fever, nausea, and vomiting, only partially controlled by antibiotics.

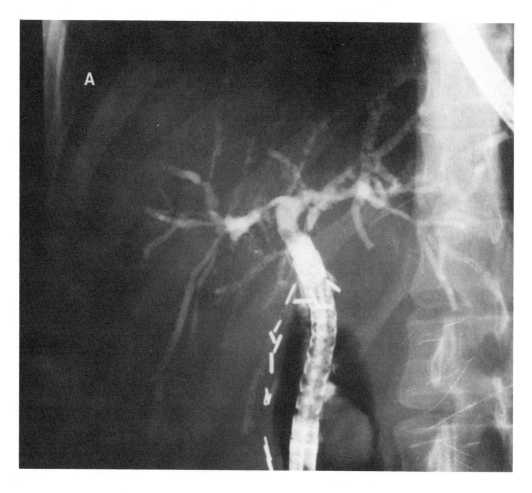

Case 46, Radiograph A

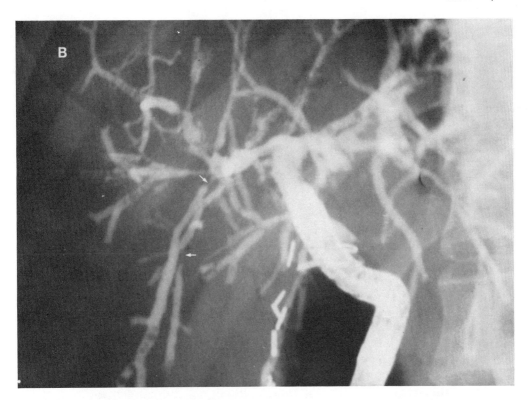

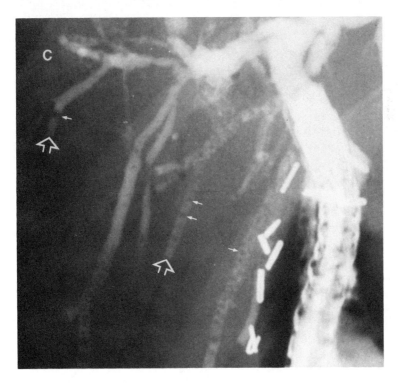

Case 46, Radiographs B & C

## Radiographic Findings

**A.**
**B.**
**&**
**C.**
The endoscope is located in the proximal common hepatic duct. The biliary radicals are grossly abnormal. There are multiple areas of stenosis (Radiograph B, arrows), distal to which many ducts show progressively increased caliber so that the normal tapering of the biliary tree is, in effect, reversed. There are multiple small filling defects within every portion of the ducts visualized (Radiograph C, arrows), and many ducts are occluded. (Radiograph C, open arrows.)

## Diagnosis

Sclerosing cholangitis.

## Comment

First note the strikingly unusual position of the endoscope. It was introduced in the usual way, entering a moderately dilated common bile duct via the duodenum. Because proximal filling of the biliary system was unsuccessful due to leakage of the contrast material around the catheter, it was elected to advance the endoscope more proximally in an effort to seat it firmly in the common hepatic duct.

The diagnosis of sclerosing cholangitis is made on the basis of multiple stenoses throughout the intrahepatic ductal system. Stenotic, normal caliber, and dilated segments are all encountered, often interspersed along the course of a single duct. Frequently, branches are "pruned," or there are areas of nonfilling, producing a sparse, "branchless tree" appearance. The presence of calculi is variable. Crohn's disease and ulcerative colitis are frequently associated with this condition. Also see Case 47.

This case illustrates an innovative approach to a difficult technical problem. The endoscopist was faced with technical failure because of inability to fill the proximal portions of the biliary tree due to leakage of the contrast material around the catheter. His solution, to use the endoscope itself as a mechanical obturator, is both simple and effective. It is to be emphasized that with ERCP, the diagnosis is made during performance of the procedure. No amount of careful review of the radiographs can substitute for a duct not opacified, a stenosis not identified, a calculus not seen.

There is no technical limitation on passage of the endoscope through surgical anastomoses. Osnes and Myren[1] have described a technique for performing the study in patients with Billroth II gastrectomy.

**D.** Here is another patient who had had a cholecystectomy 4 years previously, presenting currently with sweating, chills, vomiting, and abdominal pain. ERCP was performed to rule out retained calculus. This case demonstrates the appearance of multiple calculi packed within a dilated biliary system. There is a single large facetted calculus in the distal common duct (1). The individual filling defects are defined as round or facetted.

## Comment

This degree of calculus packing of the extrahepatic ducts is unusual, and the calculus filling of intrahepatic ducts in Radiographs A through C above is decidedly rare. The pattern of the contrast material overall produces a "honeycomb" appearance which is characteristic for the presence of calculi and is dependent on each calculus being surrounded by a small amount of contrast material.

## Reference

1. Osnes M, Myren J: Endoscopic retrograde cholangio-pancreatography (ERCP) in patients with Billroth II partial gastrectomies. *Endoscopy* 7:227-232, 1975.

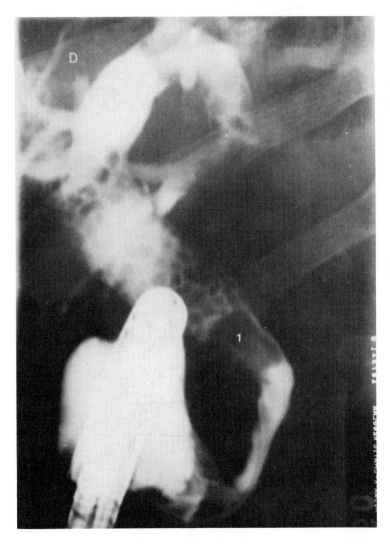

**Case 46, Radiograph D**

## CASE 47
### (Accompanying Radiographs A-D)

**Clinical History**

A 30-year-old white male with a 6-month history of watery diarrhea diagnosed as ulcerative colitis on biopsy. Evaluated now for persistently abnormal liver function studies. No jaundice, pruritus or clinical evidence of hepatitis.

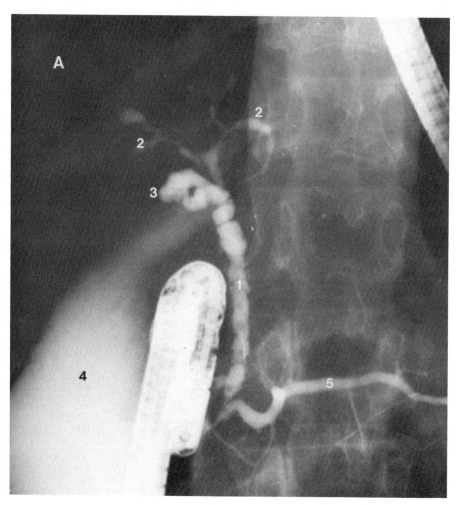

**Case 47, Radiograph A**

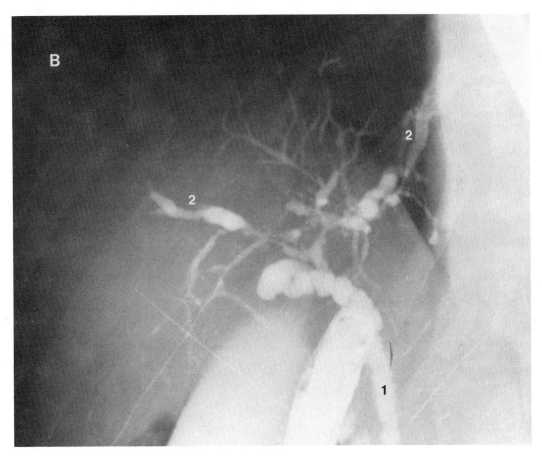

Case 47, Radiograph B

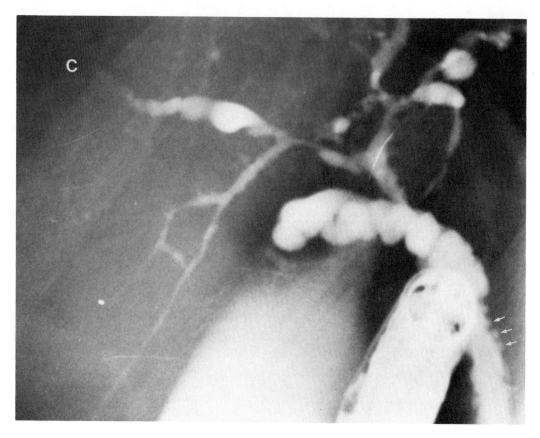

Case 47, Radiograph C

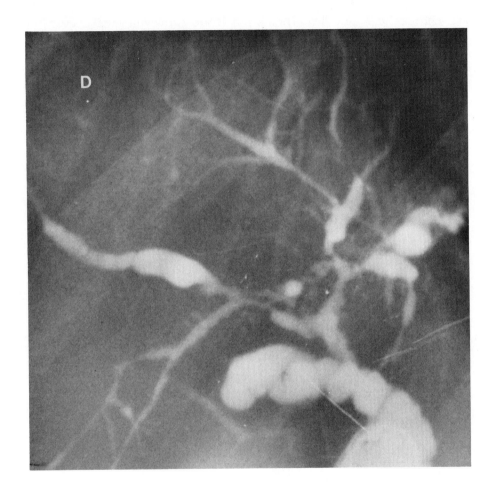

Case 47, Radiograph D

## Radiographic Findings

A. The common bile duct (1) shows an irregular and uniquely beaded margin which
& has a nearly regular saccular pattern. Intrahepatic branches (2) are sparse, and
B. large areas of the biliary system are not filled. Short-segment stenoses pre-
dominate. Duct occlusions are also present in addition to areas of proximal duct
dilatation.

C. The features just described for A and B are shown in greater detail. The sac-
& culated appearance of the common bile duct margin is especially well visual-
D. ized in Radiograph C (arrows) while Radiograph D shows the striking caliber
variation of intrahepatic branches. Note in Radiograph A that the cystic duct (3),
gallbladder (4), and pancreatic duct (5) all appear normal.

## Comment

A more typical case of sclerosing cholangitis would be difficult to find. The strikingly abnormal appearance of the common duct is in marked contrast to the normal
cystic duct and gallbladder. The overall pattern of involvement produces what is
a characteristic ERCP appearance so that there is essentially no differential. The
intrahepatic findings alone may be mimicked by cirrhosis in that the latter produces
nonfilled or thinned, stretched ducts, but without alternating narrowing and dilatation. Cholangiocarcinoma produces an identical appearance except for the common
duct changes seen here. Note that the radiologic features alone do not permit
distinction between sclerosing cholangitis occurring as a primary condition and the
form which occurs secondary to Crohn's disease or ulcerative colitis.

In cases such as this, the overall capacity of the intrahepatic ductal system is
reduced and care should be taken not to overdistend those segments that are capable
of being filled. Caution is the best guide for injection pressure and volume.

# NOTES

## CASE 48
### (Accompanying Radiographs A-D)

**Clinical History**

A 66-year-old male with 40-pound weight loss over the previous 4 months, abnormal liver function studies, and hyperbilirubinemia.

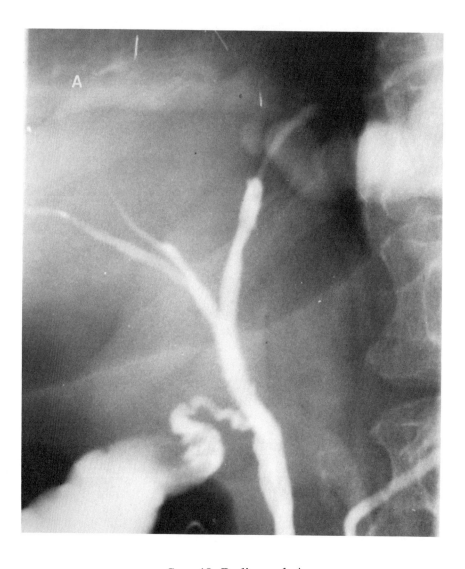

Case 48, Radiograph A

Case 48, Radiograph B

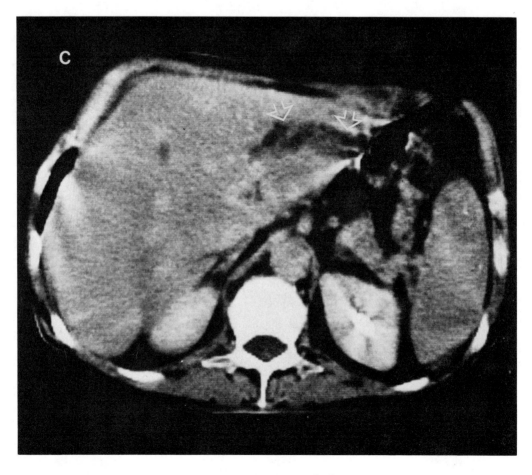

Case 48, Radiograph C

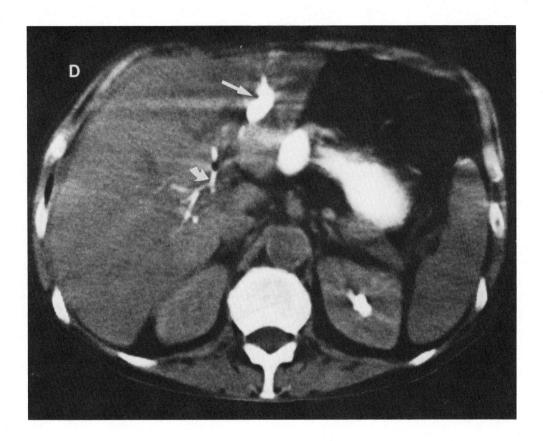

Case 48, Radiograph D

**Radiographic Findings**

A. There is a localized stenosis in a branch of the left hepatic duct about 4 cm
&  proximal to the common hepatic junction. In association with this area is a
B. cluster of lobulated, somewhat rounded structures resembling a bunch of grapes, although in portions they are somewhat saccular. The right intrahepatic ducts are normal as are the entire extrahepatic biliary system and pancreatic duct.

C. CT scan showing dilated tubular structures in the left lobe of the liver (open arrows) which do not enhance following intravenous injection of contrast material and are, therefore, consistent with dilated biliary radicals. No dilatation of right lobe radicals is noted.

D. CT scan following attempted transhepatic cholangiogram, confirms the duct dilatation in the left lobe (arrow) and normal caliber within the right lobe (curved arrow).

**Diagnosis**

Localized intrahepatic bile duct stenosis, probable carcinoma.

**Comment**

This case demonstrates an extremely localized involvement of a segment of the left lobe of the liver resulting in severe obstruction of a branch duct. In the absence of an identifiable calculus or of a previous history of hepatic trauma or surgery, malignant stenosis is considered probable. Even though surgical confirmation is lacking in this case, the diagnosis is considered definitive for the characteristic features demonstrated (see Case 49).

NOTES

# CASE 49
## (Accompanying Radiographs A-E)

### Clinical History

A 57-year-old white male who had had a cholecystectomy for cholelithiasis 13 years previously. Two years previously he had had episodes of shaking chills, fever, and darkened urine occurring over a 1-year period. Papillotomy was performed 1 year prior to current examination and three retained calculi in the common duct were removed. The patient was asymptomatic for the following year, when symptoms reappeared which included epigastric pain, pruritus, chills, and fever accompanied by dark urine. ERCP was performed to exclude retained calculus.

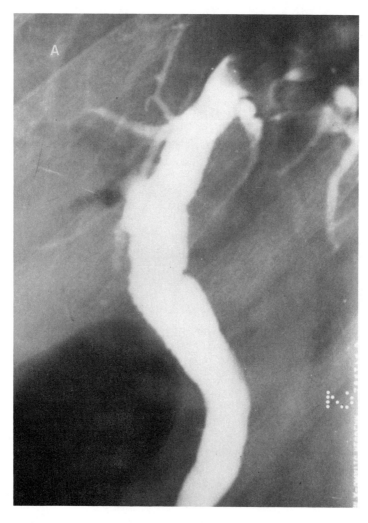

**Case 49, Radiograph A**

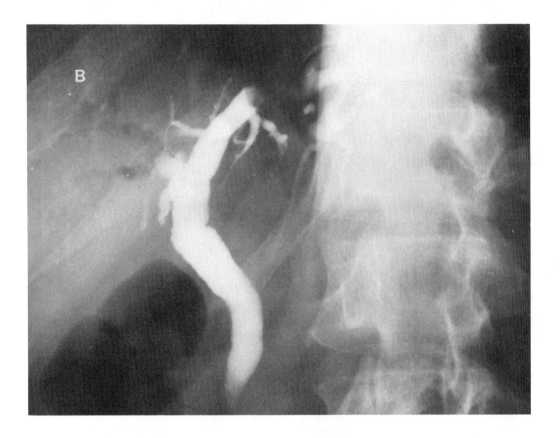

Case 49, Radiograph B

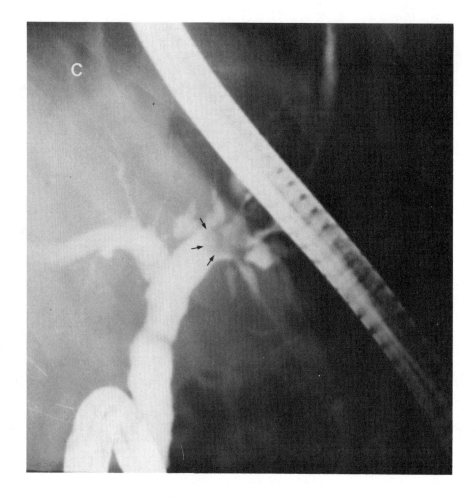

Case 49, Radiograph C

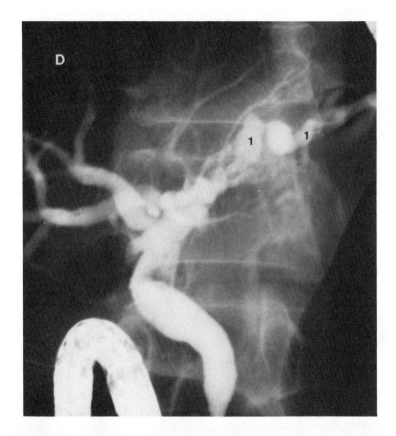

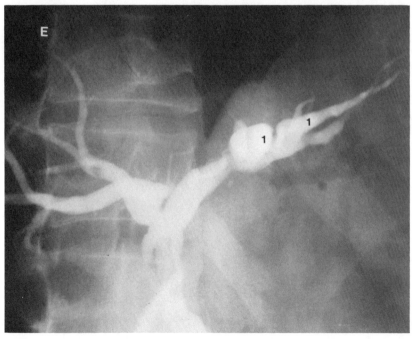

Case 49, Radiographs D & E

## Radiographic Findings

A. The extrahepatic biliary system is moderately dilated. There is filling of a few
& intrahepatic radicals which are normal in caliber on the right. On the left there is
B. a convex filling defect in the left main hepatic duct about 4 cm proximal to the
common hepatic duct. This shows a well-defined margin and forms an acute
angle with the duct laterally. No proximal duct filling is identified.

C. Reexamination 5 months later shows a similar appearance of the filling defect
D. initially (Radiograph C, arrows) followed by outlining of a dilated, somewhat
& saccular proximal left hepatic duct (Radiographs D and E, 1).
E.

## Diagnosis

Obstruction of the left hepatic duct by calculus, with proximal duct dilatation.

## Comment

The filling defect in the left hepatic duct is quite characteristic for calculus (compare to Case 44 A and B), and is important to distinguish from the misleading effect of overlying small branches.

The similarities between Cases 48 and 49 are striking as both demonstrate a localized abnormality of a single segment of the biliary tree. In both, saccular dilatation of a branch duct is present proximal to a near complete obstruction. Isolated duct dilatation should always be regarded as an indication of the presence of gross pathology since it is the final event, the outcome of disease processes which may include calculus or neoplasm. Thus, ERCP should not cease at the point of demonstration of the dilatation alone, but must continue until a point of obstruction is defined which, by characteristic shape, differentiates the disease process present.

# Index by Case Number

# INDEX

## V.  Carcinoma

## VI.  Self-Evaluation